MACHINISTS LIBRARY
Volume III

Toolmakers Handy Book

by Rex Miller

THEODORE AUDEL & CO.
a division of
THE BOBBS-MERRILL CO., INC.
Indianapolis/New York

FOURTH EDITION
FIRST PRINTING

Published by The Bobbs-Merrill Company, Inc. Indianapolis/New York
Manufactured in the United States of America

Library of Congress Cataloging in Publication Data

Miller, Rex, 1929-
    Machinists library.

    Revision of the 3rd ed. of three volumes previously published separately in
1978 under titles: Machinists library, basic machine shop, Machinists library:
machine shop, and Machinists library: toolmakers handy book.
    Includes indexes.
    Contents: v. 1. Basic machine shop—v. 2. Machine shop,—v. 3. Toolmakers
handy book.
    1. Machine-shop practice.   I. Title.
TJ1160.M566   1984   621'.9'02   83-6383
ISBN 0-672-23381-9 (v.1)
ISBN 0-672-23382-7 (v.2)
ISBN 0-672-23383-5 (v.3)
ISBN 0-672-23380-0 (set)

Foreword

The purpose of this book, in conjunction with the other two of this series, is to aid in providing a better understanding of the fundamental theory and operations needed by those persons who desire to become toolmakers and diemakers. The beginning student or apprentice must possess an adequate knowledge of the basic principles of layout, design of jigs and fixtures, and operation of centerless grinders and lapping machines if he desires to become capable of performing the various toolmaking and diemaking operations. He must also understand gear theory, helix calculations, cam design, and heat-treating operations.

Calculations and illustrations have been used generously to make the book understandable to both students and workers. Instructors should find the material helpful in presenting the how-to-do-it phase of toolmaking and diemaking jobs.

The basic fundamental theories of heat-treating and the processes of annealing, hardening, and tempering of metals are presented in an easy-to-understand manner. Diagrams and illustrations of ovens and methods of controlling temperatures inside the ovens are discussed thoroughly for both heat-treating and cold-treating operations. Furnace brazing is also dealt with in a thorough manner.

An individual who is ambitious to improve his knowledge of toolmaking and diemaking will find the material very helpful—whether he is studying alone in his spare time or is an apprentice working under close supervision on the job. This book is presented at a time when there is a definite trend toward expanded opportunities for vocational training of young adults and continuing education for anyone with the interest and desire.

Rex Miller

Acknowledgments

A number of companies have been responsible for furnishing illustrative materials and procedures used in this book. At this time, the author and publisher would like to thank them for their contributions.

Browne & Sharpe Manufacturing Company
Cincinnati Milacron Company
DoAll Company
Federal Products
The Heald Machine Company
A. F. Holden Company
Illinois Gear Company
Machinery's Handbook, Industrial Press
Johnson Gas Appliance Company
Lepel Corporation
Norton Company
Paul and Beekman, Inc.
L. S. Starrett Company
Thermolyn Corporation

We would also like to thank Dr. Ellsworth Russell for his contributions to the third edition of this book. Dr. Russell died December 19, 1980. He lived a long and productive life as an educator and machinist.

Contents

CHAPTER 6

CHAPTER 7

CHAPTER 8

CHAPTER 9

CHAPTER 10

CHAPTER 11

CHAPTER 1

Layout Work

In machine shop operations, *layout* means the marking or scribing of lines on a piece of metal to indicate in full scale the area that is to be machined. The working lines and centers, as given on the blueprint, are reproduced on the casting or forging so that the machinist can machine the workpiece properly.

Layout work is simply "machine shop mechanical drawing." It is an operation that requires skill and precision; on some pieces of work there is only a small margin of metal to allow for any inaccuracies.

FUNDAMENTAL GEOMETRIC CONSTRUCTION

A few geometrical constructions are basic in preliminary layout work. A knowledge of these can be extremely helpful to the mechanic. The following problems can be helpful:

Problem 1. To erect a perpendicular to a straight line from any point above the line.

From a point above the line (Fig. 1-1), use a radius that is long enough to cut the given line at points *F* and *G*; from these points describe arcs intersecting at the point *E*. Place a straightedge on points *A* and *E*, and draw the line from *A* perpendicular to the line *GF*.

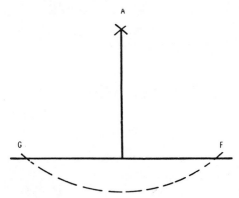

Fig. 1-1. Erecting a perpendicular to a straight line from any point above the line.

Problem 2. To erect a perpendicular at the end of a given line.

From any center point *F* above the line *BC* (Fig. 1-2), describe a circle through the given point *A* on the line, intersecting the given line at point *D*; draw line *DF*, extending it to intersect the circle at point *E*. Then draw the perpendicular *AE*.

Problem 3. To bisect a given line.

Using the extremities (points *A* and *B*) of the given line (Fig. 1-3) as centers and a radius larger than one-half the total length of the line *AB*, describe arcs that intersect at points *C* and *D*. The line *CD* bisects line *AB*. Therefore, line *AE* is equal to line *EB*.

Problem 4. To divide a given line into any number of equal parts.

10

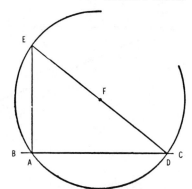

ig. 1-2. Erecting a perpendicular
at the end of a given line.

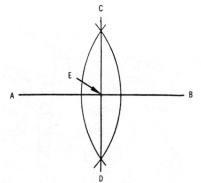

Fig. 1-3. Bisecting a given line.

Two methods can be used to divide a given line into any number of equal parts. In assuming that the given line *AB* (Fig. 1-4) is to be divided into five equal parts, draw the diagonal line *AC* and mark off five units of equal length. Draw the line *B5* and then construct lines *a1, b2,* etc., parallel to line *B5.* The line *AB* will then be divided into five equal parts.

Another method can be used to divide a given line into any number of equal parts. A scale, a triangle, and a T-square can be used to divide a given line *MS* into seven equal parts, as shown in Fig. 1-5. First, erect the perpendicular *SY.* Place the "zero" mark of the scale on *M* and place the seventh unit of the scale to coincide with the line *SY.* Construct line *MF,* drawing it lightly, and mark the units (inch) divisions as shown. Then, using the triangle and T-square, construct perpendiculars from the points on line *MF,*

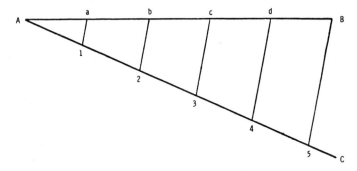

Fig. 1-4. Dividing a given line into any number of equal parts.

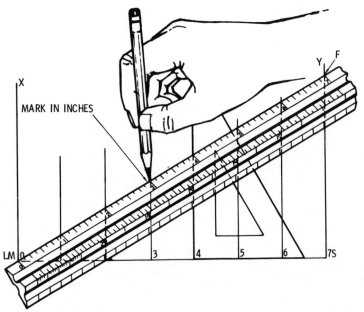

Fig. 1-5. A second method of dividing a given line into any number of equal parts.

intersecting the line *MS* at the points *1, 2, 3, 4, 5, 6*, dividing it into seven equal parts.

Problem 5. To divide a given line into proportional parts.

Two methods can be used to divide a line into proportional

12

parts. Assume the given line *AB* (Fig. 1-6A) is to be divided into three parts proportional to *1*, *3*, and *5*. Draw the diagonal line *AC* and select a scale that gives 9 equal units. Set the zero mark of the scale on *A* and mark off on *AC* the first, fourth, and ninth units. Draw a line from the ninth unit to *B* and lines parallel to this line from the mark at the first and fourth unit. Line *AB* is now divided proportionally.

Another method is to erect a perpendicular line at *B* (Fig. 1-6B). Select a scale as before and place the zero mark at *A*. Slide the other end of the scale up until unit 9 coincides with the perpendicular line at point *C*. Draw line *AC* and mark first and fourth units on the line. Draw perpendicular lines from these marks to line *AB*. Line *AB* is now divided proportionally. Any proportional division will work in the same manner.

Problem 6. To construct a line through a given point parallel to a given line.

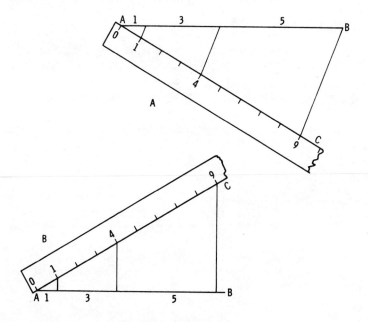

Fig. 1-6. Dividing a given line into proportional parts.

13

Using the given point C (Fig. 1-7) as a center point, describe an arc tangent to the given line AB; the radius is then equal to the distance from the given point C to the given line AB. Using any point B on the line AB remote from point A as a center point, and the same radius, describe an arc. Draw a line through the point C tangent to the arc; then the line CD is parallel to the given line AB.

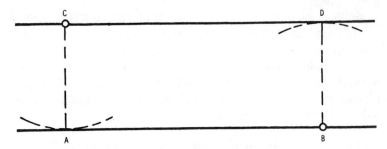

Fig. 1-7. Constructing a line through a given point parallel to a given line.

Problem 7. To construct a square on a given line.

Erect a perpendicular to the given line AB (Fig. 1-8), and lay off a distance AD equal to AB. Using the points D and B as center points and a radius equal to AB, describe arcs that intersect, as at point E. Draw the lines BE and DE to complete the square.

Problem 8. To construct a 45° angle at a given point on a given line.

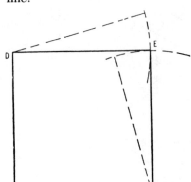

Fig. 1-8. Constructing a square on a given line.

14

At any point D on the given line *AB* (Fig. 1-9), erect a perpendicular *DG*. The point *C* is the given point on line *AB*. Using the point *D* as the center point and radius *DC*, describe the arc *CFE*, cutting the perpendicular at point *E*. Construct the line *CE;* then angle *ECB* is 45 degrees.

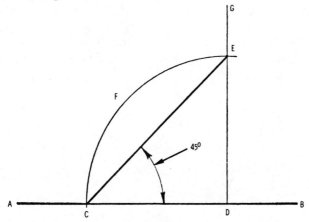

Fig. 1-9. Constructing an angle of 45° at a given point on a given line.

Problem 9. To construct a 60° angle at a given point on a given line.

At any point *D* on the given line *AB* (Fig. 1-10), strike arc *EF* using any convenient radius. With *F* as the center and the same

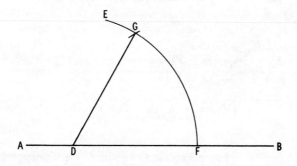

Fig. 1-10. Constructing a 60° angle at a given point on a given line.

15

radius, strike an arc intersecting the first arc at *G*. Draw *DG*. Then *GDB* is a 60° angle. To construct a 30° angle, bisect angle *GDB*.

Problem 10. To bisect an angle.

In Fig. 1-11, the given angle *ACB* can be bisected by describing an arc having the center point *C*. Using the points *A* and *B* at which the arc cuts the sides of the angle as center points, describe equal arcs, intersecting at point *D*. The line *CD* divides the angle into two equal parts.

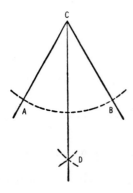

Fig. 1-11. Bisecting an angle.

Problem 11. To locate the center of a circle.

In Fig. 1-12, draw any chord *MS*. Using the points *M* and *S* as center points and any reasonable radius, describe arcs *L, F, N,* and *G;* construct a line through the intersections giving the diameter *AB* of the circle. Repeating the construction, using the points *A* and *B* as center points, describe the arcs *E, H, J,* and *K.* A line drawn through the intersections of these arcs intersect the line *AB* at the point *O*, which is the center point of the circle.

Problem 12. To describe an arc that is tangent to two lines that are perpendicular.

In Fig. 1-13, the two given lines *OX* and *OY* are perpendicular. Using the given radius *r* and the point *O* as a center point, describe arcs *a* and *b*, intersecting lines *OY* and *OX* at points *A* and *B*,

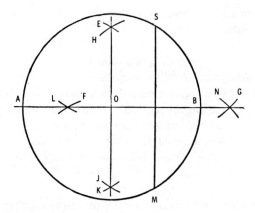

Fig. 1-12. Locating the center point of a circle.

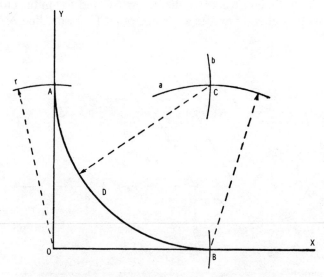

Fig. 1-13. One method of describing an arc that is tangent to two lines that are perpendicular to each other.

respectively. Using the same radius r and points A and B as center points, describe arcs a and b, intersecting at point C. Then, with the same radius r and point C as the center point, describe the arc ADB, which is tangent to the perpendicular lines OY and OX at points A and B, respectively.

17

Problem 13. To describe an arc that is tangent to two lines at acute or obtuse angles.

In Fig. 1-14A, the two given lines *OX* and *OY* are at an acute angle at *(A)* and at an obtuse angle at *(B)*. Using the given radius of the arc *R* and two convenient points *M* on each line, scribe the arcs as shown. Draw lines tangent to the two arcs with centers on the original lines. These tangent lines are parallel to the original lines and will intersect at point *N*. From point *N*, erect perpendiculars *NS* and *NT* to lines *OX* and *OY*. Using point *N* as the center, draw arc *ST*, which is tangent to lines *OX* and *OY*.

Problem 14. To inscribe a pentagon inside a circle.

In Fig. 1-15, construct two diameters *AC* and *BD* in the circle, intersecting at right angles at center point *O*. Bisect line *AO* to

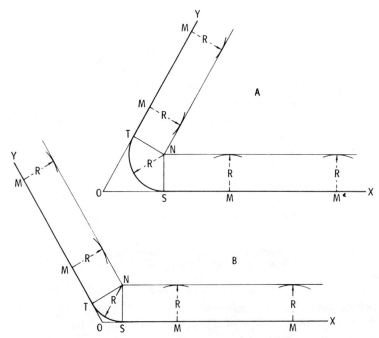

Fig. 1-14. One method of describing an arc that is tangent to two lines at acute or obtuse angles.

locate point E; with radius EB and point E as the center point, locate point F on line AC. With radius BF and point B as center, locate points G and H on the circumference. Using the same radius BF, step around the circumference to locate points I and K; join the points so located to form the pentagon.

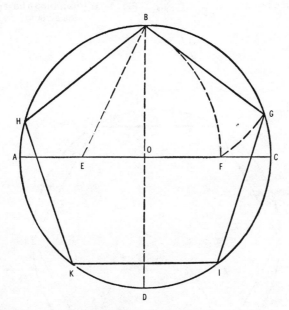

Fig. 1-15. Inscribing a pentagon inside a circle.

Problem 15. To inscribe a hexagon inside a circle.

Draw the diameter AB of the circle (Fig. 1-16). Using the radius AC and points A and B as center points, describe arcs that intersect the circumference at points G, F, D, and E. Draw lines AD, DE, etc., to complete the hexagon. The points D, E, etc., can be located by using the dividers to step off the radius around the circle, but this method is less accurate.

Problem 16. To construct a hexagon on a given line.

From the ends of the given line AB (Fig. 1-17), using the radius AB, describe arcs that intersect at point H; from the point H, with

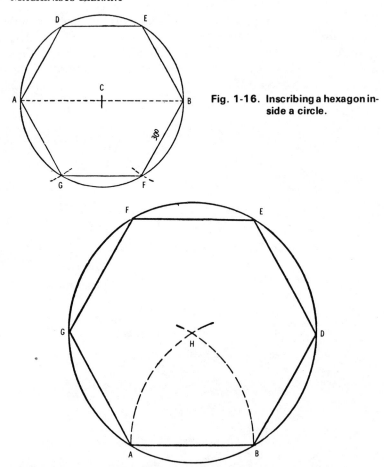

Fig. 1-16. Inscribing a hexagon inside a circle.

Fig. 1-17. Constructing a hexagon on a given side.

the radius *AB*, describe a circle. Using the same radius, locate the points *G*, *F*, etc. Join these points to complete the hexagon.

Problem 17. To solve for the length of a side of an inscribed square.

In Fig. 1-18, the diameter *AC* in which the square *ABCD* is inscribed is equal to the hypotenuse of the right triangle *ABC*.

20

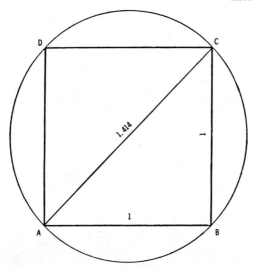

Fig. 1-18. One method of solving for the length of a side of an inscribed square.

$$AC = \sqrt{(1)^2 + (1)^2} = \sqrt{2} = 1.414$$

Example: What is the size of the largest square jaw that can be machined from a piece of 3-inch round stock?

Substituting in the formula, the side of the jaw can be calculated as follows:

$$\text{Side } AB = \frac{3}{1.414} = 2.121 \text{ inches}$$

Problem 18. To find the distance between the opposite sides of a hexagon.

In the right triangle (ABC in Fig. 1-19), let the radius AC of the circumscribed circle be equal to 1 (unity). Then side $AB = \frac{1}{2}$, and the length of side BC can be calculated:

$$\text{Side } BC = \sqrt{(1)^2 - (0.5)^2} = \sqrt{0.75} = 0.866$$

As the value 0.866 is the ratio between one-half the total distance BD (or BC) and the radius AC of the circumscribed circle, it is also

21

the ratio between the total distance *BD* across the sides of the hexagon and the diameter *AE* of the circumscribed circle, or

$$BD = AE \times 0.866$$

Example: What is the size of the largest hexagon that can be machined from a piece of 3-inch round stock?

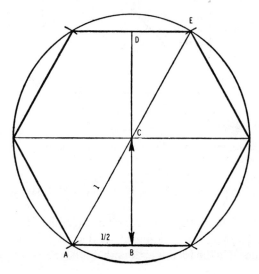

Fig. 1-19. One method of solving for the distance between the opposite sides of a hexagon.

Substituting in the formula, the total distance between the opposite sides of the hexagon can be calculated as follows:

$$BD = 3 \times 0.866 = 2.598 \text{ inches}$$

Thus, the milling cutters must be set 2.598 inches apart to mill the largest hexagon that can be cut from a piece of 3-inch round stock.

The foregoing problems are typical of the various problems that can be encountered in layout work. The student should master them before beginning practice in laying out.

LAYING OUT

A mechanic with a working knowledge of the foregoing fundamental geometrical constructions is prepared for exercises that embrace the problems usually encountered in laying out. In production work in large shops, where numerous jigs and fixtures locate the work and guide the cutting tools, repeated laying out is unnecessary. However, especially in small shops without such equipment, the work must be laid out. Laying out can be done more accurately on a surface that is finished than on a rough surface.

In general, complete layout is not made in a single operation, but continues progressively because some of the lines and center points can be located more accurately on a finished surface. Moreover, if the complete layout were made at first, some of the lines could become indistinct while handling the workpiece, or be removed by turning or planing tools. Furthermore, it is not always practical to make a complete layout in a single operation because some lines can be located more accurately from a finished surface.

Before laying out a machine part, it is important to consider its particular requirements in order to obtain a definite idea of the relation that the various surfaces must bear to one another when the part is finished and assembled. As the method of procedure is governed somewhat by the subsequent machining operations, it is advisable to determine the most practical method of machining before beginning to lay out the work.

In order to do layout work it is necessary to be able to read the drawings that supply the needed information. It is assumed that the reader has some knowledge and experience with drawings and with the English inch system of measurement. However, another system is being introduced that must be considered.

Metric Measurement

Rapid technological development has emphasized the need for a single worldwide coordinated measurement system. Such a system was suggested over 300 years ago. However, the wide distribution of manufactured material in the past few decades has brought increased recognition of the need for such a system. Most countries of the world, the major exception being the United States, are on or are committed to using the metric system. Many industries in the United States, General Motors being an example, are in the process of making the change. It is important that the reader become familiar with this system. However, at present American progress toward metrification has halted.

The General Conference on Weights and Measures gave the revised metric system its new title Système International d'Unites with the universal abbreviation SI units. These basic units are given in Table 1-1. The unit of length is the meter, which is much too large for practical use for the toolmaker. Therefore, the millimeter, abbreviated mm, $\frac{1}{1000}$ th (0.001) of a meter, has been accepted as the standard for the precision manufacturing industries (Table 1-2). The metric system is based on units that are divided or multiplied by powers of ten. Each power of ten has a prefix and a symbol, as indicated in Table 1-1.

The toolmaker may come into contact, during this conversion period, with several methods of using the metric system on drawings. Some drawings are made in inches and are are provided with a conversion chart to give each measurement in metric. Some are drawn in metric and have a conversion chart into inches. Some have dual measurements—that is, both the metric and inch measurement are given for each dimension. In such drawings it is customary to enclose one set in brackets to avoid confusion. Most organizations are encouraging their people to "think metric." The general practice is to design, draw, and manufacture using only metric measurements.

Fortunately, it is not difficult to "think metric." SI units of measure in the metric system conform to reason, are consistent, and fit together. The toolmaker will be faced with machines that have metric measurements. The measuring tools he uses will have metric scales. The drawings provided will be in millimeters. Ratios and angular measurements are the same in both systems. A number of differences should be kept in mind while using metric drawings. These include:

1. The word *metric* will appear in the largest-size lettering.
2. The symbol for first angle projection or third angle projection (Fig. 1-20) will appear in the title block.

Table 1-1. The International System of Units

BASIC SI UNITS

Quantity	Name of Unit	Symbol
Length	Meter	m
Mass (weight)	Kilogram	kg
Time	Second	s
Electric current	Ampere	A
Temperature	Kelvin	K
Luminous intensity	Candela	cd
Amount of substance	Mole	mol

SUPPLEMENTARY UNITS

Plane angle	Radian	rod
Solid angle	Steradian	sr

SI NUMERICAL PREFIXES

Select the appropriate prefix which will give a numerical value within a range of 0.1 to 1000. Double prefixes should be avoided. Conventional practices in some industries will often dictate the prefix to be used. For example, the garment industry will most frequently use centimeters, while engineering drawings will be in millimeters.

Prefix	Symbol	Pronunciation	Multiplication Factor	Meaning
exa	E	ex'a	$1\ 000\ 000\ 000\ 000\ 000\ 000 = 10^{18}$	One quintillion times
peta	P	pet'a	$1\ 000\ 000\ 000\ 000\ 000 = 10^{15}$	One quadrillion times
tera	T	ter'a	$1\ 000\ 000\ 000\ 000 = 10^{12}$	One trillion times
giga	G	ji'ga	$1\ 000\ 000\ 000 = 10^{9}$	One billion times
mega	M	meg'a	$1\ 000\ 000 = 10^{6}$	One million times
kilo	k	kil'o	$1\ 000 = 10^{3}$	One thousand times
hecto	h	hek'toe	$100 = 10^{2}$	One hundred times
deka	da	dek'a	$10 = 10^{1}$	Ten times
unit			$1 = 10^{0}$	One
deci	d	des'i	$0.1 = 10^{-1}$	One tenth of
centi	c	sen'ti	$0.01 = 10^{-2}$	One hundredth of
milli	m	mil'i	$0.001 = 10^{-3}$	One thousandth of
micro	μ	mi'kro	$0.000\ 001 = 10^{-6}$	One millionth of
nano	n	nan'o	$0.000\ 000\ 001 = 10^{-9}$	One billionth of
pico	p	peek'o	$0.000\ 000\ 000\ 001 = 10^{-12}$	One trillionth of
femto	f	fem'toe	$0.000\ 000\ 000\ 000\ 001 = 10^{-15}$	One quadrillionth of
atto	a	at'to	$0.000\ 000\ 000\ 000\ 000\ 001 = 10^{-18}$	One quintillionth of

Table 1-2. Metric Units by Type of Industry

Industry	Unit of Measure	Symbol
Topographical	kilometer	km
Building, Construction	meter	m
Lumber, Cabinetmaking	centimeter	cm
Mechanical Design, Manufacturing	millimeter	mm

3. A note will read, "Unless otherwise specified, dimensions are in millimeters."
4. Fasteners, screw threads, and drills will be in metric sizes (Tables 1-3, 1-4, and 1-5).
5. A symbol θ in front of a dimension indicates it is a diameter.
6. A comma is frequently used instead of a decimal point.
7. A comma is never used to denote thousands. A space may be used for such a purpose.
8. Millimeter measurement is based on a one-place decimal system; two- and three-place decimals are used only for critical tolerances.
9. A millimeter value of less than one will have a zero to the left of the decimal mark.
10. A limit dimension will have the same number of places to the right of the decimal mark.

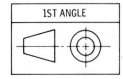

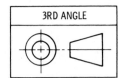

Fig. 1-20. Projection symbols.

Table 1-3. ISO Metric Thread—Fine Series

Diameter	8	10	12	14	16	18	20
Pitch	1.0	1.25	1.25	1.5	1.5	1.5	1.5
Basic effective diameter	7.350	9.188	11.188	13.026	15.026	17.026	19.026
Depth of thread in screw. . . .	0.61	0.77	0.77	0.92	0.92	0.92	0.92
Area of Root diameter (mm²)	36.0	56.3	86.0	116	157	205	259
Diameter of tapping drill	7.0	8.8	10.8	12.5	14.5	16.5	18.5
Diameter	22	24	30	36	42	48	
Pitch	1.5	2.0	2.0	3.0	3.0	3.0	
Basic effective diameter	21.026	22.701	28.701	34.051	40.051	46.051	
Depth of thread in screw. . . .	0.92	1.23	1.23	1.84	1.84	1.84	
Area of Root diameter (mm²)	319	365	586	820	1210	1540	
Diameter of tapping drill	20.5	22.0	28.0	33.0	39.0	45.0	

Table 1-4. ISO Metric Thread—Coarse Series

Diameter	2	2.5	3	4	5	6	8	10
Pitch	0.4	0.45	0.5	0.7	0.8	1.0	1.25	1.5
Basic effective diameter	1.740	2.208	2.675	3.545	4.480	5.350	7.188	9.026
Depth of thread in screw	0.25	0.28	0.31	0.43	0.49	0.61	0.77	0.92
Area of Root diameter (mm^2) . . .	1.79	2.98	4.47	7.75	12.7	17.9	32.8	52.3
Diameter of tapping drill	1.6	2.05	2.5	3.3	4.2	5.0	6.8	8.5
Diameter	12	16	20	24	30	36	42	48
Pitch	1.75	2.0	2.5	3.0	3.5	4.0	4.5	5.0
Basic effective diameter	10.863	14.701	18.376	22.051	27.727	33.402	39.077	44.752
Depth of thread in screw	1.07	1.23	1.53	1.84	2.15	2.45	2.76	3.07
Area of Root diameter (mm^2) . . .	76.2	144	225	324	519	759	1050	1380
Diameter of tapping drill	10.2	14.0	17.5	21.0	26.5	32.0	37.5	43.0

Table 1-5. Twist Drill Data

Metric Drill Sizes (mm)[1]		Decimal Equiv-alent in Inches (Ref)	Metric Drill Sizes (mm)[1]		Decimal Equiv-alent in Inches (Ref)
Preferred	Available		Preferred	Available	
	0.40	0.0157	1.20		0.0472
	0.42	0.0165	1.25		0.0492
	0.45	0.0177	1.30		0.0512
	0.48	0.0189		1.35	0.0531
0.50		0.0197	1.40		0.0551
	0.52	0.0205		1.45	0.0571
0.55		0.0217	1.50		0.0591
	0.58	0.0228		1.55	0.0610
0.60		0.0236	1.60		0.0630
	0.62	0.0244		1.65	0.0650

[1]Metric drill sizes listed in the "Preferred" column are based on the R′40 series of preferred numbers shown in the ISO Standard R497. Those listed in the "Available" column are based on the R80 series from the same document.

Table 1-5. Twist Drill Data (Continued)

Metric Drill Sizes (mm)[1]		Decimal Equiv-alent in Inches (Ref)	Metric Drill Sizes (mm)[1]		Decimal Equiv-alent in Inches (Ref)
Preferred	Available		Preferred	Available	
0.65		0.0256	1.70		0.0669
	0.68	0.0268		1.75	0.0689
0.70		0.0276	1.80		0.0709
	0.72	0.0283		1.85	0.0728
0.75		0.0295	1.90		0.0748
	0.78	0.0307		1.95	0.0768
0.80		0.0315	2.00		0.0787
	0.82	0.0323		2.05	0.0807
0.85		0.0335	2.10		0.0827
	0.88	0.0346		2.15	0.0846
0.90		0.0354	2.20		0.0866
	0.92	0.0362		2.30	0.0906
0.95		0.0374	2.40		0.0945
	0.98	0.0386	2.50		0.0984
1.00		0.0394	2.60		0.1024
	1.03	0.0406		2.70	0.1063
1.05		0.0413	2.80		0.1102
	1.08	0.0425		2.90	0.1142
1.10		0.0433	3.00		0.1181
	1.15	0.0453		3.10	0.1220
3.20		0.1260		7.80	0.3071
	3.30	0.1299	8.00		0.3150
3.40		0.1339		8.20	0.3228
	3.50	0.1378	8.50		0.3346
3.60		0.1417		8.80	0.3465
	3.70	0.1457	9.00		0.3543
3.80		0.1496		9.20	0.3622
	3.90	0.1535	9.50		0.3740
4.00		0.1575		9.80	0.3858
	4.10	0.1614	10.00		0.3937
4.20		0.1654		10.30	0.4055
	4.40	0.1732	10.50		0.4134
4.50		0.1772		10.80	0.4252
	4.60	0.1811	11.00		0.4331
4.80		0.1890		11.50	0.4528
5.00		0.1969	12.00		
	5.20	0.2047	12.50		0.4724
5.30		0.2087	13.00		0.4921

[1]Metric drill sizes listed in the "Preferred" column are based on the R'40 series of preferred numbers shown in the ISO Standard R497. Those listed in the "Available" column are based on the R80 series from the same document.

Table 1-5. Twist Drill Data (Continued)

Metric Drill Sizes (mm)[1]		Decimal Equivalent in Inches (Ref)	Metric Drill Sizes (mm)[1]		Decimal Equivalent in Inches (Ref)
Preferred	Available		Preferred	Available	
	5.40	0.2126		13.50	0.5118
5.60		0.2205	14.00		0.5315
	5.80	0.2283			0.5512
6.00		0.2362		14.50	0.5709
	6.20	0.2441	15.00		0.5906
6.30		0.2480		15.50	0.6102
	6.50	0.2559	16.00		0.6299
6.70		0.2638		16.50	0.6496
	6.80[2]	0.2677	17.00		0.6693
	6.90	0.2717		17.50	0.6890
7.10		0.2795	18.00		0.7087
	7.30	0.2874		18.50	0.7283
7.50		0.2953	19.00		0.7480
	19.50	0.7677		37.00	1.4567
20.00		0.7874	38.00		1.4961
	20.50	0.8071		39.00	1.5354
21.00		0.8268	40.00		1.5748
	21.50	0.8465		41.00	1.6142
22.00		0.8661	42.00		1.6535
	23.00	0.9055		43.50	1.7126
24.00		0.9449	45.00		1.7717
25.00		0.9843		46.50	1.8307
26.00		1.0236	48.00		1.8898
	27.00	1.0630	50.00		1.9685
28.00		1.1024		51.50	2.0276
	29.00	1.1417	53.00		2.0866
30.00		1.1811		54.00	2.1260
	31.00	1.2205	56.00		2.2047
32.00		1.2598		58.00	2.2835
	33.00	1.2992	60.00		2.3622
34.00		1.3386			
	35.00	1.3780			
36.00		1.4173			

[1]Metric drill sizes listed in the "Preferred" column are based on the R'40 series of preferred numbers shown in the ISO Standard R497. Those listed in the "Available" column are based on the R80 series from the same document.

[2]Recommended only for use as a tap drill size.

Equipment

The first essential requirement for laying out the work is a flat surface for supporting the work and the layout instruments. This can be a surface plate, platen (reciprocating table of a planer), drill press table, etc. The surface must be true—that is, flat. This modern space age demands precision measuring. The more precise the measuring, the more important it is to have a true surface plate. Some surface plates, usually made of granite, are manufactured to a flatness tolerance measured in millionths of an inch or in ten-thousandths of a millimeter. Some of the necessary layout tools are: surface gage, combination square, V-blocks, scriber, center punch, prick punch, and machinists' hammer. Other tools might include gage blocks, sine bars, indicator gages, vernier calipers, and vernier height gages.

Surface Preparation

A coating material should be used to make lines scribed on the metal stand out clearly. This colored coating material is applied to the surface of the workpiece and causes the form outlined by the scriber lines to stand out in sharp contrast to the background. The coating material should not wash or rub off; the solution used depends on the kind of metal in the workpiece.

Several different kinds of *layout blue* or *layout dope* are available commercially. The material is inexpensive, dries quickly, and produces clear-cut layout lines.

Chalk is often used on rough castings, but it has the disadvantage of being easily rubbed off the metal. For laying out large castings or a large quantity of small pieces, whiting mixed with alcohol or water is often used. When mixed with alcohol, whiting dries quicker and does not rust the surface.

For many years a copper sulfate solution called blue vitriol has been used as a coating for machined iron or steel surfaces. The surface to be scribed should be clean before applying the solution, which is painted on. A thin copper plating is produced on the surface; fine lines can be seen easily because of the difference in color between the copper and the metal underneath.

Aluminum and copper materials can be coated easily with vemilion and shellac thinned with alcohol. This treatment produces a surface on which layout lines can be scribed without danger of scratching the underlying metal. Use alcohol as a wash for cleaning the coating substance from the metal.

Methods of Locating Center Points

Several methods of locating center points can be used. One of the first operations for the machinist is to locate center points on a piece of round stock that is to be turned in a lathe. Hand methods of locating center points are generally used, but a centering machine is sometimes available in large shops.

Hermaphrodite calipers—After the surface has been coated, the hermaphrodite calipers should be adjusted to a radius slightly larger than the radius of the stock; describe the arcs $a, b, c,$ and d so that each pair is approximately at right angles (Fig. 1-21). Then draw the diagonals XX and YY through the intersections of the arcs. The intersection of the diagonals is the approximate center point. Center punch the intersection lightly, and repeat the operation for the other end of the stock.

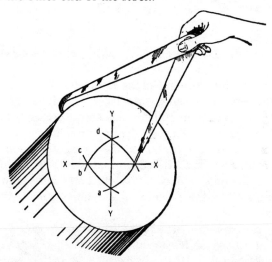

Fig. 1-21. Using the hermaphrodite calipers to locate the center point on round stock.

Dividers—In the absence of other instruments, dividers can be used to locate center points on round stock, but this is usually regarded as a makeshift method (Fig. 1-22). Coat the surface of the metal and place the stock on a surface plate. Set the dividers for a distance slightly larger than the radius of the stock, and scribe two pairs of lines at approximate right angles for each end of the

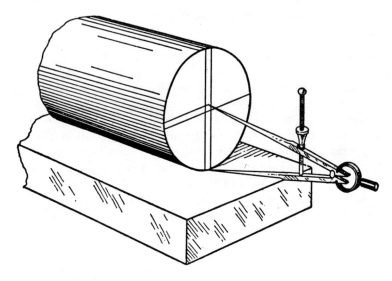

Fig. 1-22. Using the dividers to locate the center point on round stock.

stock. The point of intersection of the diagonals of the small square indicates the center point of the stock.

Surface gage—Place the stock to be centered in a V-block on a surface plate (Fig. 1-23). Then set the scriber of the surface gage at an elevation slightly off center; scribe four lines on the stock, turning it a quarter turn after each line is scribed. Repeat the operation on the other end of the stock. The center point of the stock is indicated by the point of intersection of the diagonals of the small square.

Combination square—In combination with the center head, the combination square blade can be used to locate center points on round stock (Fig. 1-24). Attach the center head to the combination square blade, place the head against the stock, and scribe a line along the edge of the blade (see *A* in Fig. 1-24). Then place the square in another position (at least one-sixth turn) and scribe another (see *B* in Fig. 1-24). The intersection of the two lines is the center point of the stock; sometimes it is desirable to check the work by scribing a third line.

Bell-cup center punch—This tool provides the quickest means of locating the center point on small round stock where precision is

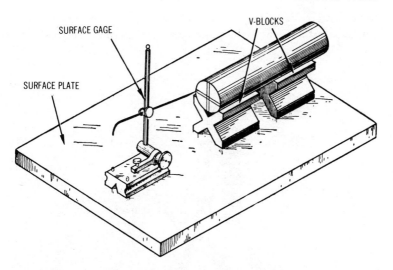

Fig. 1-23. Using the surface gage to locate the center point on round stock.

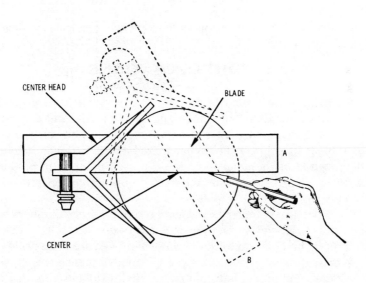

Fig. 1-24. Using the combination square and center head to locate the center point on round stock.

not required. Varying diameters of stock can be center punched. For greatest accuracy, the punch must be aligned "square" with the work—not tilted at an angle (Fig. 1-25).

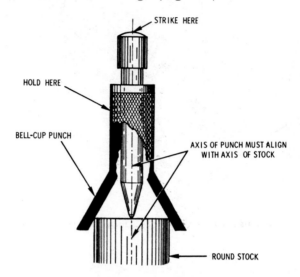

Fig. 1-25. Using the bell-cup punch to locate the center point on round stock.

LAYOUT OPERATIONS

Holes must be drilled in the proper places if parts are to fit together as intended. Unless a drill jig is to be used, careful layout is required in locating holes to be drilled. Various means can be used to locate properly holes that are to be drilled, depending on the location of the holes in relation to each other or to certain surfaces. Of course, the surface must be properly prepared and coated.

Pipe flanges, cylinder heads, etc., require various numbers of bolt holes, depending on their size and required service. When four to six holes, or multiples of these numbers, are to be drilled, two diameters are usually drawn at right angles; these intersect the "bolt circle" at four equidistant points, dividing it into four quadrants. For multiples of four bolt holes each quadrant can be subdivided further.

For devices that have six holes, the intersection of each end of one of the diameters can be used as a center point (*YY* in Fig. 1-26), and with the radius of the bolt circle set on the dividers, arcs are described to intersect the bolt circle at points *1, 2,* etc. Thus, four bolt-hole center points are located, and the center points from which the arcs are described can be used as the center points of the other two bolt holes.

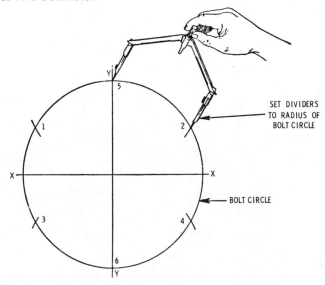

Fig. 1-26. Methods of laying out center points for bolt holes.

Pipe Flange

A metal strip can be placed across the base of the flange and the combination square and surface gage used to locate the center for laying out (Fig. 1-27). Place the work on V-blocks, and adjust with shims, if necessary, until the work is level. Set the surface gage at the same height as the center point and scribe diameter *XX*. Using the combination square, scribe a second diameter *YY* at right angles to the first diameter *XX*. Then scribe the bolt circle with the dividers. If the flange layout is for eight bolts, for example, set the dividers at one-half the quadrant measured on the bolt circle; with center points at *a* and *b* describe intersecting arcs, giving four bolt-hole centers at points *1, 2, 3,* and *4.* The other bolt-hole center

35

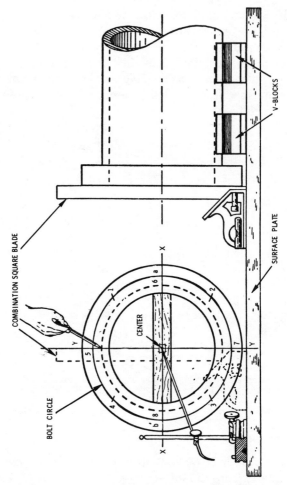

Fig. 1-27. Method of locating bolt-hole center points in layout of a pipe flange.

points 5, 6, 7, and 8 are located at the intersections of the axes with the bolt circle. The prick punch can be used to mark the center points of the bolt holes as they are located.

Valve Cage

Piston valves usually work in bushings that are pressed into the cylinder casting (Fig. 1-28). To lay out the work from working

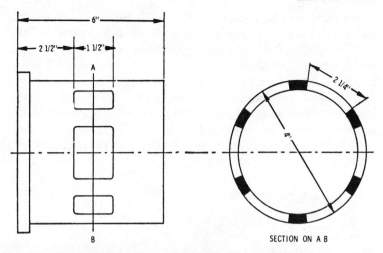

Fig. 1-28. Working drawing of a valve cage or bushing, giving the dimensions required for layout.

drawings, first locate the center (*C* in Fig. 1-29) on the end without the shoulder, and scribe a circle near the outer edge. Then, with the work resting on V-blocks, by means of the surface gage and the combination square, scribe the two axes (*XX* and *YY*) at right angles.

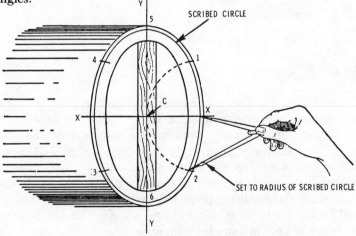

Fig. 1-29. Locating the center point and spacing for center lines on the scribed circle in laying out a valve bushing.

37

For the six port openings (see Fig. 1-28), set the dividers to the radius of the scribed circle, and with the intersections of the scribed circle and the axis XX as center points scribe arcs *1, 2, 3,* and *4*. The intersections of the YY axis and the scribed circle are locations of the other points *5* and *6* (see Fig. 1-29).

Set the surface gage at the elevation of the center C, and turn the bushing around until one point (*4*, for example) registers with the scriber of the surface gage; then scribe the line AB (Fig. 1-30) and another line on the opposite side of the bushing. These lines are the center lines for two of the port openings.

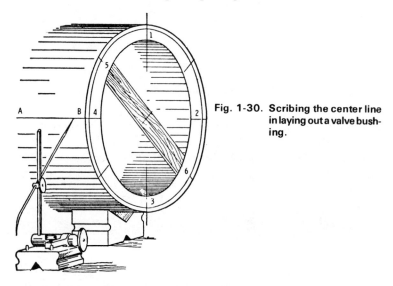

Fig. 1-30. Scribing the center line in laying out a valve bushing.

Referring to the drawing or blueprint (see Fig. 1-28), note that the width of the port opening is 2.25 inches. To lay out the sides, set the surface gage from the combination square at one-half the width (or 1.125 inch) lower, and scribe line CD (Fig. 1-31) and a similar line on the opposite side of the bushing.

Then set the surface gage 2.25 inches higher and scribe line EF and a similar line on the opposite side (see Fig. 1-31). The lines CD and EF are marks for the sides of the port. The sides of the other ports are laid out in a similar manner.

In Fig. 1-32, the lines AB, CD, and EF correspond to the lines lettered similarly in Fig. 1-31. To lay out the top and bottom lines

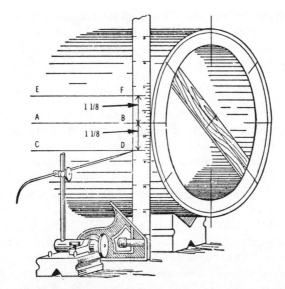

Fig. 1-31. Scribing the port side lines in laying out a valve bushing.

of the ports, place the bushing in a vertical position on the surface plate. The dimensions can be obtained from the working drawings (see Fig. 1-28); here the top of the port is 4 inches from the shoulder end of the bushing, and the bottom is 2.50 inches from the shoulder end of the piece.

Set the scriber at 2.50 inches above the surface plate (position *1* in Fig. 1-32), and scribe line *GH*. Similarly, obtain line *IJ* (posi-

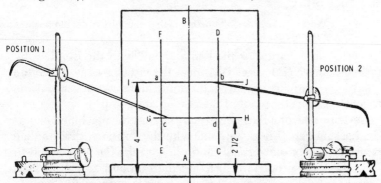

Fig. 1-32. Scribing the port top and bottom lines in laying out a valve bushing.

tion 2); thus, the layout *a b c d* is completed for one of the ports. In the actual layout, of course, the line *GH* can be continued around the bushing in one operation for use on all the ports; line *LJ* can be used in the same manner.

Main Bearing Cap

The layout of a main bearing cap can be used to illustrate, among other things, the use of an angle plate and a surface plate to hold the work in a vertical position. From the working drawings (Fig. 1-33), note the dimensions necessary for laying out the holes.

As shown in Fig. 1-33, place a metal block flush with the machined end. Place the bearing cap on the surface plate, and set the surface gage at an elevation of 0.250 inch—this is the height of the center of the bearing above the lug ends (see Fig. 1-33). With the surface gage at this setting, scribe the line *AB* across the metal block; the center point is on this line. Measure across the bearing or lugs and mark one-half this distance on line *AB* to locate the center point of the bearing.

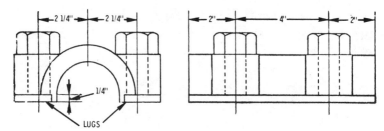

Fig. 1-33. Working drawings of a main bearing cap, giving the dimensions required for laying out.

Check the location of the center point with the dividers by scribing an arc (*L* in Fig. 1-34) near the circular edge. Place the combination square on the center point, and scribe a vertical line to locate the center point of the casting at the top.

Rest one side of a thin steel square against the machined end of the bearing cap (Fig. 1-35), and with the 2.25-inch division line registering with the vertical center line, scribe the lines *A* and *B*. Using the same procedure, turn the square around, and scribe lines *C* and *D* at a distance of 2.25 inches from the center line.

Place the cap on its machined end on a surface plate, and clamp

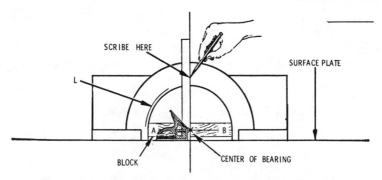

Fig. 1-34. Locating the center point and scribing the center line laying out a main bearing cap.

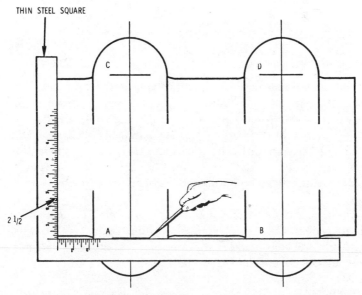

Fig. 1-35. Scribing the longitudinal lines for locating the bolt holes in layout of the main bearing cap.

it to an angle plate (Fig. 1-36). As indicated in the working drawings (see Fig. 1-33), the bolt-hole centers are located 2 inches from the cap ends.

Set the surface gage at 2 inches from the surface plate (position *I* in Fig. 1-36), and scribe the lines *E* and *F*. The intersections of these lines with lines *C* and *A* (scribed in Fig. 1-35) are the locations

41

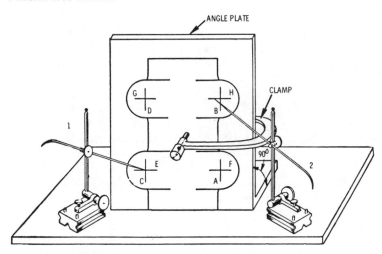

Fig. 1-36. Scribing the transverse lines for locating the bolt holes in layout of the main bearing cap.

of the center points for the bolt holes at one end of the bearing cap.

From the working drawings (see Fig. 1-33), it can be noted that the other two bolt-hole centers are 6 inches (2 + 4) from the end of the cap. Set the surface gage at an elevation of 6 inches (position 2 in Fig. 1-36), and scribe the lines G and H. Then the intersections with lines D and B (scribed in Fig. 1-35) are the locations of the center points of the bolt holes for the second end of the bearing cap.

The machined end of the bearing cap resting on the surface plate insures its being in a vertical position. If the cap is not true vertically, the lines E and F, for example, cannot be scribed at the same distance from the end of the cap.

Crank Arm

In laying out a crank arm, first locate the center point at the shaft end by means of a metal strip and combination square. The center point can be checked with the dividers (Fig. 1-37). Then the crank arm can be placed on its side on the surface plate. The crank arm should be leveled and the central axis ABCD scribed, as shown in Fig. 1-38. Set the surface gage at the center of the shaft end (position 1); adjust the leveling block until the center of the crank

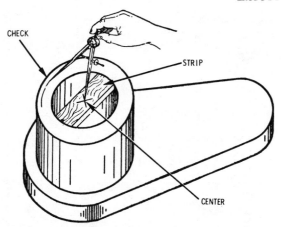

Fig. 1-37. Locating the shaft center point in layout of a crank arm.

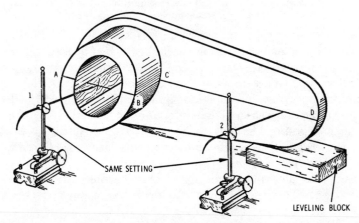

Fig. 1-38. Leveling the arm and scribing the center axis in layout of the crank arm.

pin end is at the same height, and scribe the axis *ABCD*. The center point of the crank pin will be located on this axis (position 2).

The next step in laying out the crank arm is to place it in a vertical position, turning it until the line *AB* registers with the blade of the combination square, and clamping it to the face of the angle plate (Fig. 1-39). The distance from the center point of the shaft end to the center point of the crank end can be obtained from the working drawing of the crank arm.

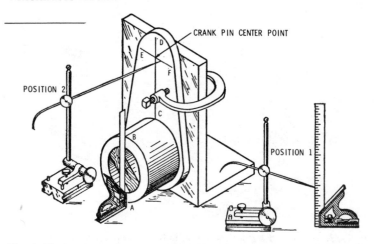

CRANK PIN CENTER POINT

POSITION 2

POSITION 1

Fig. 1-39. Locating the crank-pin center point in layout of the crank arm. The arm is held in vertical position by an angle plate.

Using the original setting of the surface gage (position *1* in Fig. 1-38), the center point of the shaft should be checked for the setting in the vertical position (see Fig. 1-39), making any minute adjustments that are necessary and checking the height with the combination square (position *1* in Fig. 1-39). The distance from the center point of the shaft to the center point of the crank pin (taken from the working drawing) can be added to this height to give the correct setting on the combination square for location of the crank pin center. Using this setting (position *2* in Fig. 1-39), scribe the line *EF* to locate the crank pin center point at the intersection of lines *CD* and *EF*. Then use the dividers to describe the crank pin circle.

Crosshead

The method for laying out a crosshead depends on the design of the crosshead—there are several kinds of crossheads. The kind of crosshead selected for illustration of layout procedures is shown in the working drawing (Fig. 1-40). For the sake of simplicity, only the details and dimensions needed for layout are given in the working drawing. No attempt is made here to explain the purpose of a crosshead, as most engine mechanics are familiar with them.

The first step in laying out the crosshead is to place metal strips across the holes for the piston rod and the wrist pin for location of

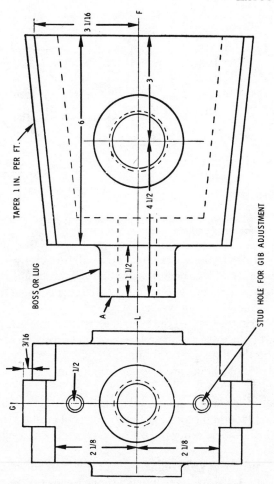

Fig. 1-40. Working drawing for a crosshead, giving the dimensions necessary for laying out.

their center points, as shown in Fig. 1-41. The center of the wrist-pin hole should be located at a correct distance from the crosshead end A, that is, 4.50 inches (as given in the working drawing). The metal should extend far enough to allow for a light cut at end A (see Fig. 1-40); therefore, the distance (4.50 inches) from the wrist-pin center point should terminate just short of the end of the casting, rather than at the end of the casting, so that the distance

45

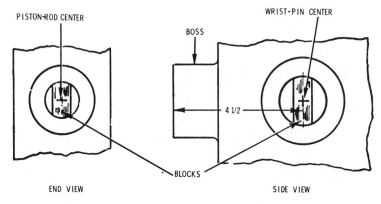

PISTON-ROD CENTER

WRIST-PIN CENTER

BOSS

4 1/2

BLOCKS

END VIEW

SIDE VIEW

Fig. 1-41. End and side views of the crosshead, showing the location of the piston-rod and wrist-pin centers.

from the end A to the wrist-pin center will be exactly 4.50 inches after the light cut has been taken at A. A line should be scribed completely around the boss to indicate to the machinist the correct amount of metal to remove at A.

As shown in the view of the crosshead from the end opposite the boss head (Fig. 1-42), place a block of metal flush with the end, and scribe the diagonals AB and CD. The intersection of these diagonals should be located on the longitudinal axis LF (see Fig. 1-40) of the casting.

Further steps in layout are shown in Fig. 1-43. The casting should be placed on blocks so that it does not contact the surface plate, using a block under the boss and two pairs of wedges at the other end.

Set the surface gage to register with the piston-rod center (position A in Fig. 1-43). Move the surface gage to position B, and adjust the wedges until the surface gage registers with the wrist-pin center. Check the wrist-pin center on the other side, and adjust the wedges on that side until the wrist-pin center point on that side also registers with the surface gage. Check the work by moving the surface gage to position C; if the centers have been located correctly, the surface gage should register with the center point on that end. This gives a three-point support and should hold the work firmly in position.

After properly locating the crosshead on the block and wedges,

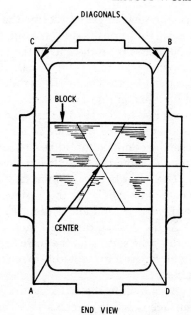

Fig. 1-42. End view of the crosshead from the end opposite the bosshead, showing the location of the center point.

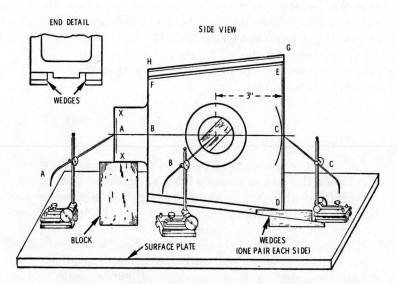

Fig. 1-43. Blocking and leveling the crosshead on a surface plate for scribing the layout lines in laying out the crosshead.

use the surface gage to scribe the line *ABC* and a corresponding line on the opposite side (see Fig. 1-43). Locate the boss line *XX* at 1.50 inches from the end of the crosshead. Note on the working drawing (see Fig. 1-40) that the center point of the wrist pin is located exactly 3 inches from the end of the machined casting (see dotted line in Fig. 1-43).

Set the dividers to 3 inches, and using the wrist-pin center point as a center point, describe an arc intersecting line *ABC* at point *C*. Use the combination square to scribe the vertical line *DCE*. The point *C* should be exactly 6 inches from point *B*.

The taper begins at point *E*, which is 3.0625 inches above the center line *ABC*. Set this distance (*C* + 3.0625 inches) on the surface gage, and locate point *E*.

As the desired taper is 1 inch per foot, or 0.5 inch in 6 inches, reduce the setting on the surface gage by 0.5 inch, and locate the point *F*. Use a straightedge to scribe line *EF* to indicate the correct taper.

As shown in the working drawing (see Fig. 1-40), the raised surface *G* is 0.1875 inch above the tapered surface on each side. Then the last setting of the surface gage can be increased 0.1875 inch to locate point *H*. Increase the setting still another 0.5 inch, and locate point *G*. Then scribe line *HG*.

The two lines *EF* and *GH* indicate the limits to which the tapered surfaces are to be machined. Repeat these operations for the other tapered surfaces.

SUMMARY

Layout work is simply machine shop mechanical drawings. Mechanical drawing or layout work is an operation that requires skill and precision. The working lines given on a blueprint are reproduced on the casting or forging so that the machinist can machine the workpiece properly.

For production work in large machine shops, where numerous jigs and fixtures locate the work and guide the cutting tools, repeatedly laying out is not necessary. However, especially in small shops without such equipment, the work must be laid out.

Holes must be drilled in the proper places if parts are to fit together as intended. Unless a drill jig is to be used, careful layout is required in locating holes to be drilled. Many ways are used to properly locate these holes.

The essential requirements for laying out the work is a flat surface for supporting the work and layout instruments. Some of the necessary layout instruments are surface gage, combination square, V-block, scriber, center punch, prick punch, and machinist's hammer.

REVIEW QUESTIONS

1. What is layout work?
2. Name a few essential tools needed for layout work.
3. What is layout blue or layout dope? Why is it used?
4. What is a bell-cup center punch?
5. What is crosshead in working drawings?
6. How is the combination square used to find the center of a piece of round stock?
7. How are the dividers used to divide a bolt circle for six equally spaced bolts?
8. How is the center of a hole or a pipe located for layout work?
9. Why are number drills obsolete in the United Kingdom?

CHAPTER 2

Jigs and Fixtures

Jigs and fixtures are devices used to facilitate production work, making interchangeable pieces of work possible at a savings in cost of production. Both terms are frequently used incorrectly in shops. A jig is a *guiding* device and a fixture a *holding* device.

Jigs and fixtures are used to locate and hold the work that is to be machined. These devices are provided with attachments for guiding, setting, and supporting the tools in such a manner that all the workpieces produced in a given jig or fixture will be exactly alike in every way.

The employment of unskilled labor is possible when jigs and fixtures can be used in production work. The repetitive layout and setup, which are time-consuming activities and require considerable skill, are eliminated. Also, the use of these devices can result in such a degree of accuracy that workpieces can be assembled with a minimum amount of fitting.

A jig or fixture can be designed for a particular job. The form to be used depends on the shape and requirement of the workpiece to be machined.

JIGS

The two types of jigs that are in general use are: (1) clamp jig and (2) box jig. A few fundamental forms of jigs will be shown to illustrate the design and application of jigs. Various names are applied to jigs, such as drilling, reaming, and tapping—according to the operation to be performed.

Clamp Jig

This device derives its name from the fact that it usually resembles some form of clamp. It is adapted for use on workpieces on which the axes of all the holes that are to be drilled are parallel.

Clamp jigs are sometimes called *open jigs*. A simple example of a clamp jig is a design for drilling holes that are all of the same size—for example, the stud holes in a cylinder head (Fig. 2-1).

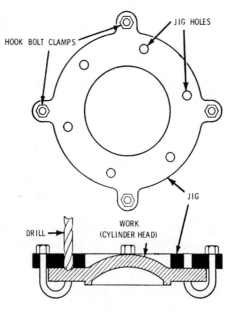

Fig. 2-1. A plain ring-type clamp jig without bushings.

As shown in the illustration, the jig consists of a ring with four lugs for clamping and is frequently called a "ring jig"; it is attached to the cylinder head and held by U-bolt clamps. When used as a guide for the drill in the drilling operation, the jig makes certain that the holes are in the correct locations because the holes in the jig were located originally with precision; therefore, laying out is not necessary.

A disadvantage of the simple clamp jig is that only holes of a single size can be drilled. Either *fixed* or *removable* bushings can be used to overcome this disadvantage. Fixed bushings are sometimes used because they are made of hardened steel, which reduces wear. Removable bushings are used when drills of different sizes are to be used or when the drilled holes are to be finished by reaming or tapping.

A *bushed clamp jig* is illustrated in Fig. 2-2. In drilling a hole for a stud, it is evident that the drill (tap drill) must be smaller in size

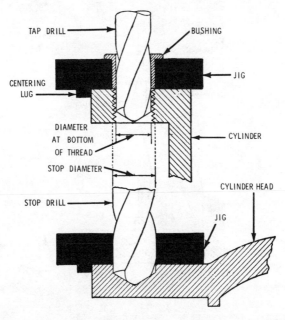

Fig. 2-2. A clamp jig, with the tap drill guided by a bushing, designed for drilling holes in the cylinder (top); the operation for a hole for the cylinder head (bottom).

than the diameter of the stud. Accordingly, two sizes of twist drills are required in drilling holes for studs; the smaller drill (or tap drill) and a drill slightly larger than the diameter of the stud are required for drilling the holes in the cylinder head. A bushing can be used to guide the tap drill.

The jig is clamped to the work after it has been centered on the cylinder and head so that the axes of the holes register correctly. Various provisions, such as stops, are used to aid in centering the jig correctly. The jig shown in Fig. 2-2 is constructed with four lugs as a part of the jig. As the jig is machined, the inner sides of the lugs are turned to a diameter that will permit the lugs to barely slip over the flange when the jig is applied to the work.

A *reversible clamp jig* is shown in Fig. 2-3. The distinguished

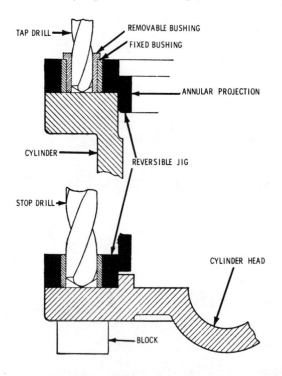

Fig. 2-3. Note the use of a reversible clamp jig for the tap drill operation (top), and reversing the jig to drill the hole for the stud in the cylinder head (bottom).

feature of this type of jig is the method of centering the jig on the cylinder and head. The position of the jig for drilling the cylinder is shown at the top of the illustration (see Fig. 2-3). An annular projection on the jig fits closely into the counterbore of the cylinder to locate the jig concentrically with the cylinder bore.

The jig is reversed for drilling the cylinder head; that is, the opposite side is placed so that the counterbore or circular recessed part of the jig fits over the annular projection of the cylinder head (see illustration at bottom of Fig. 2-3).

This type of jig is often held in position by inserting an accurately fitted pin through the jig and into the first hole drilled. The pin prevents the jig turning with respect to the cylinder as other holes are drilled.

A simple jig that has locating screws for positioning the work is shown in Fig. 2-4. The locating screws are placed in such a way that the clamping points are opposite the bearing points on the work. Two setscrews are used on the long side of the work; but in this instance, as the work is relatively short and stiff, a single lug and setscrew (see B in Fig. 2-4) is sufficient.

This is frequently called a "plate jig" since it usually consists of only a plate, which contains the drill bushings, and a simple means of clamping the work in the jig, or the jig to the work. Where the jig is clamped to the work, it sometimes is called a "clamp-on jig."

"Diameter jigs" provide a simple means of locating a drilled

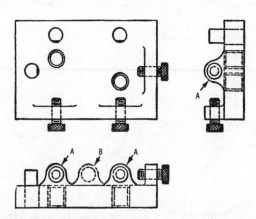

Fig. 2-4. A simple jig that uses locating screws to position the work.

hole exactly on a diameter of a cylindrical or spherical piece (see Fig. 2-5).

Another simple clamp jig is called a "channel jig" and derives its name from the cross-sectional shape of the main member, as shown in Fig. 2-6. They can be used only with parts having fairly simple shapes.

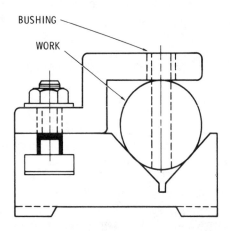

Fig. 2-5. Diameter jig.

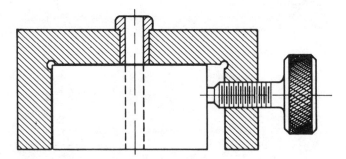

Fig. 2-6. Channel jig.

Box Jig

Box jigs (sometimes called "closed jigs") usually resemble a boxlike structure. They can be used where holes are to be drilled in

56

the work at various angles. A design of box jig that is suitable for drilling the required holes in an engine link is shown in Fig. 2-7. The jig is built in the form of a partly open slot in which the link is moved up against a stop and then clamped with the clamp bolts A, B, and C.

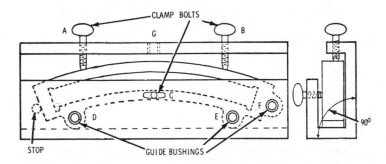

Fig. 2-7. Using the box jig for drilling holes in an engine link.

The bushings D and E guide the drill for drilling the eccentric rod connections, and the bushing F guides the drill for the reach rod connections. The final hole, the hole for lubrication at the top of the link, is drilled by turning the jig 90°, placing the drill in the bushing G.

This type of jig is relatively expensive to make by machining, but the cost can be reduced by welding construction, using plate metal. In production work, the pieces can be set and released quickly.

A box jig with a hinged cover or leaf, which may be opened to permit the work to be inserted and then closed to clamp the work into position, is usually called a "leaf jig" (see Fig. 2-8). Drill bushings are usually located in the leaf. However, bushings may be located in other surfaces to permit the jig to be used for drilling holes on more than one side of the work. Such a jig, which requires turning to permit work on more than one side, is known as a "rollover jig."

A box jig for angular drilling (Fig. 2-9) is easily designed by providing the jig with legs of unequal length, thus tilting the jig to the desired angle. This type of jig is used where one or more holes are required to be drilled at an angle with the axis of the work.

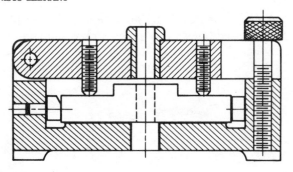

Fig. 2-8. Leaf jig.

As can be seen in the illustration (see Fig. 2-9), the holes can be drilled in the work with the twist drill in a vertical position. Sometimes the jig is mounted on an angular stand rather than providing legs of unequal length for the jig. A box jig for drilling a hole in a ball is shown in Fig. 2-10.

In some instances the work can be used as a jig (Fig. 2-11). In the illustration, a bearing and cap are used to show how the work can be arranged and used as a jig. After the cap has been planed and fitted, the bolt holes in the cap are laid out and drilled. The cap is clamped in position, and the same twist drill used for the bolt holes is used to cut a conical spot in the base. This spotting operation provides a starting point for the smaller tap drill (*A* and *B* in Fig. 2-11).

Also, both parts can be clamped together and drilled with a tap drill (*C* in Fig. 2-11). Then the tap drill can be removed, and the holes for the bolts enlarged by means of a counterbore (*D* in Fig. 2-11).

Some factors of prime importance to keep in mind with jigs are: the proper clamping of the work, support of the work while machining, and provision for chip clearance.

When excessive pressure is used in clamping, some distortion can result. If the distortion is measurable, the result is inaccuracy in final dimensions. This is illustrated in an exaggerated way in Fig. 2-12. The clamping forces should be applied in such a way that will not produce objectionable distortion.

It is also important to design the clamping force in such a way

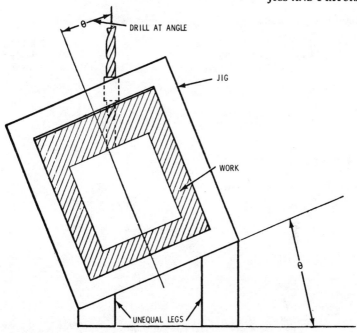

Fig. 2-9. A box jig with legs of unequal length, used for drilling holes "at an angle."

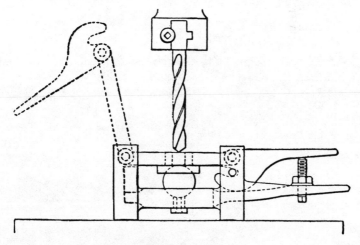

Fig. 2-10. A box jig used for drilling a hole in a ball.

59

that the work will remain in the desired position while machining. This is illustrated in Fig. 2-13.

Fig. 2-14 shows the need for the jig to provide adequate support while the work is being machined. In the example shown in Fig. 2-12, the cutting force should always act against a fixed portion

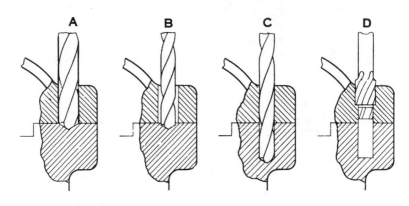

Fig. 2-11. Using the work as a jig. In (A) the same drill used for the bolt holes is used to cut a conical spot in the base. This forms a starting point for the smaller tap drill, as shown in (B). In (C), the cap and bearing are clamped together and drilled by means of a tap drill, after which the tap drill is removed and a counterbore is used to enlarge the holes for the bolts, as shown in (D).

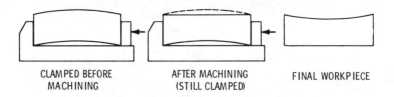

CLAMPED BEFORE MACHINING

AFTER MACHINING (STILL CLAMPED)

FINAL WORKPIECE

Fig. 2-12. Effects of excessive pressure.

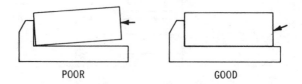

POOR

GOOD

Fig. 2-13. Effects of clamping force.

and not against a movable section. Fig. 2-13 illustrates the need to keep the points of clamping as nearly as possible in line with the cutting forces of the tool. This will reduce the tendency of these forces to pull the work from the clamping jaws. Support beneath the work is necessary to prevent the piece from distorting. Such distortion can result in inaccuracy and possibly a broken tool (see Fig. 2-14).

Adequate provision must be made for chip clearance, as illustrated in Fig. 2-15. The first problem is to prevent the chips from becoming packed around the tool. This could result in overheating and possible tool breakage. If the clearance is not great enough, the chips cannot flow away. If there is too much clearance, the bushing will not guide the tool properly.

The second factor in chip clearance is to prevent the chips from interfering with the proper seating of the work in the jig. This is illustrated in Fig. 2-16.

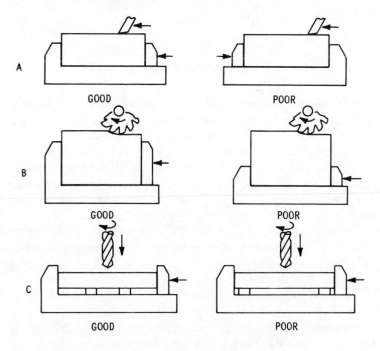

Fig. 2-14. Support for work during machining.

61

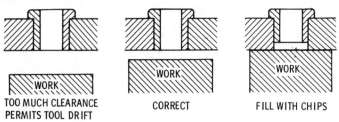

Fig. 2-15. Provision for chip clearance.

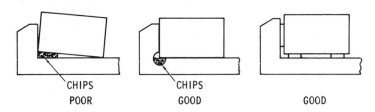

Fig. 2-16. Provision for chip clearance.

FIXTURES

As mentioned previously, a fixture is primarily a *holding* device. A fixture anchors the workpiece firmly in place for the machining operation, but it does not form a guide for the tool.

It is sometimes difficult to differentiate between a jig and a fixture, as their basic functions can overlap in the more complicated designs. The best means of differentiating between the two devices is to apply the basic definitions, as follows: (1) the jig is a *guiding* device, and (2) the fixture is a *holding* device.

A typical example of a fixture is the device designed to hold two or more locomotive cylinders in position for planing (Fig. 2-17). This fixture is used in planing the saddle surfaces. In the planing operation, two or more cylinders are placed in a single row, the fixture anchoring them firmly to the planer bed.

The fixture consists of heavy brackets or angles, with conical projections that permit the bores of the cylinders to be aligned accurately with each other. The end brackets are made with a single conical flange; the intermediate brackets are made with double conical flanges. A bolt through the center of the flanges aligns the cylinder bores when it is tightened. The legs of the 90° angle brackets at the ends are bolted firmly to the planer table. The intermediate brackets are also bolted to the planer table and aid in holding the assembly in firm alignment for the machining operation. The use of fixtures can result in a considerable saving in the time required to set the work, and they also ensure production of accurate work.

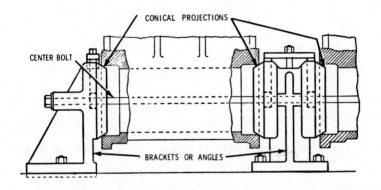

Fig. 2-17. A fixture used to hold locomotive cylinders in position for planing the surfaces of the saddles.

An indexing fixture can be used for machining operations that are to be performed in more than one plane (Fig. 2-18). It facilitates location of the given angle with a degree of precision.

A disk in the indexing fixture is held in angular position by a pin that fits into a finished hole in the angle iron and into one of the holes in the disk. The disk is clamped against the knee by a screw and washer while the cut is being taken. As the holes are properly spaced in the disk (index plate), the work attached to the disk can be rotated into any desired angular position. Radial drilling operations can be performed when a projecting plate is provided with a jig hole.

The same general principles concerning clamping, support while machining, and chip clearance as covered in jigs apply as well to fixtures.

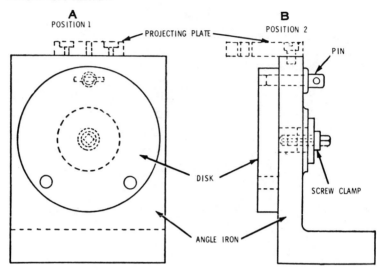

Fig. 2-18. A simple type of indexing fixture that can be used to facilitate machining at accurately spaced angles.

SUMMARY

Jigs and fixtures are devices used to locate and hold the work that is to be machined. A jig is a guiding device, and a fixture is a holding device. A jig or fixture can be designed for a particular job. The form to be used depends on the shape and requirements of the workpiece that is to be machined.

There are generally two types of jigs used: the clamp jig and the box jig. Various names are applied to jigs, such as drilling, reaming, and tapping, according to the operation to be performed. Clamp jigs are sometimes called open jigs. Frequently, jigs are named for their shape, such as plate, ring, channel, and leaf.

A fixture anchors the workpiece firmly in place for the machining operation, but it does not form a guide for the tool. It is sometimes difficult to differentiate between a jig and a fixture, as their basic functions can overlap in the more complicated designs.

It is necessary to provide proper clamping of the work, support for the work while machining, and adequate provision for chip clearance.

REVIEW QUESTIONS

1. What is a jig?
2. Name the various types of jigs.
3. What is a fixture?
4. What is an indexing fixture?
5. Explain the importance of proper clamping of the work.
6. What is meant by proper support for the work while machining?
7. What happens if no provision is made for chip clearance? If too much clearance is provided?

CHAPTER 3

Helix and Spiral Calculations

In the past, machinists have tended to use the terms *helix* and *spiral* interchangeably. Generally, in machine shop usage, the terms should not be used interchangeably; these terms should be understood by machinists, and their misuse avoided.

For general machine shop usage, the terms can be defined as follows:

1. A *helix* is a curve generated from a point that both rotates and advances axially on a cylindrical surface. The lead screw on a lathe is an example of a helix.
2. A *spiral* is a curve generated from a point that has three distinctive motions: (a) rotation about the axis; (b) advancement parallel with the axis; and (c) an increasing or decreasing distance (radius) from the axis.

When a *cylindrical* workpiece is placed between centers, on a milling machine and rotated by the index head as the table advances, a *helical groove* is milled by the cutter. When a *tapered* workpiece is placed between centers, tilted so that the top element is horizontal and then rotated by the dividing head as the table advances, a *spiral groove* is milled by the cutter. The basic difference between a helix and a spiral is illustrated in Fig. 3-1.

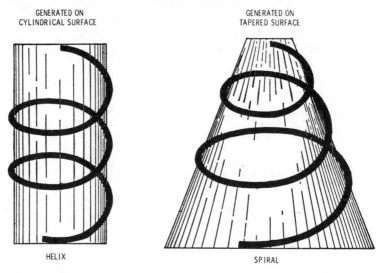

GENERATED ON
CYLINDRICAL SURFACE

GENERATED ON
TAPERED SURFACE

HELIX

SPIRAL

Fig. 3-1. Basic difference between a helix (left) and a spiral (right).

MILLING A HELIX

These are the essential requirements for milling a helix:

1. The table should be set at the correct angle.
2. The index head should be set to rotate the work in correct ratio to the table movement.
3. The work should be fed toward the cutter by the table movement.

The *pitch*, or *lead*, of a helix is the distance that the table (carrying the workpiece) travels as the work is rotated by the index head through one complete revolution (Fig. 3-2). The terms "lead" and "pitch" are identical in meaning. Pitch is probably a

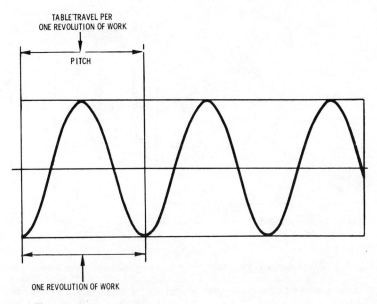

Fig. 3-2. The pitch of a helix.

more proper term; however, "lead" is more commonly used in the machine shop.

Angle of Table Swivel

This angle is the angle through which the table must be turned to cut a helix; the *table angle* is equal to the angle of the helix. Two methods can be used to determine the table angle for cutting a helix.

If a helix is laid out in a single plane, the hypotenuse of a right triangle represents the helix. The other two sides of the right triangle represent the circumference of the work and the pitch (Fig. 3-3).

The angle *AOB* (see Fig. 3-3), which is the angle of the helix, is called the table angle in the illustration because it is the angle through which the table must be turned to cut the helix correctly. If the triangle were cut out and wrapped around a cylindrical workpiece, the hypotenuse *OA*, which represents the developed helix, would coincide at all points with the helix.

69

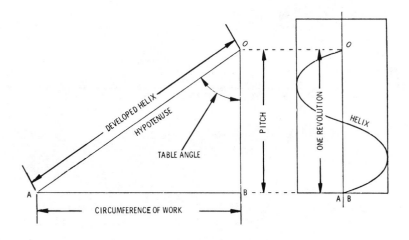

Fig. 3-3. Development of a helix by laying out, to determine the table angle.

The correct table position for cutting a helix is illustrated by angle A in Fig. 3-4. Angle A is equal to angle B, which is called the angle of the helix and is formed by the intersection of the helix and a line parallel with the axis of the work. Angle A is equal to angle B

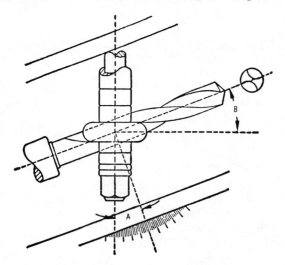

Fig. 3-4. Correct position of the table for cutting a helix.

because their corresponding sides are perpendicular. The helix angle depends on the pitch of the helix and the diameter of the work, and it varies inversely with the pitch for any given diameter.

Turning the table to an angular position for cutting the helix prevents distortion of the shape of the cut and obtains clearance for the milling cutter. The pitch of the helix is not changed by turning the table to any angular position.

Trigonometry provides a more accurate method of determining the table angle. If the pitch and circumference of the work are given, the tangent of the table angle can be found. The pitch and circumference of the work are considered to be the sides of a right triangle (Fig. 3-5). After determining the value of the tangent, the angle can be obtained from a table of natural tangents.

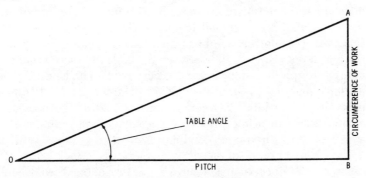

Fig. 3-5. Using trigonometry to determine the table angle.

If, in the triangle AOB in Fig. 3-5, we let the side AB equal the circumference of the work, and let the side OB equal the pitch, then:

$$\text{Tangent of the table angle} = \frac{AB \, (\text{circumference})}{OB \, (\text{pitch})}$$

Example: Determine the table angle required to cut a helix that has a pitch of 16 inches and a diameter of 4.5 inches.

The circumference of the work is equal to πD (3.141592654 × 4.5). Substituting in the formula:

$$\text{Tangent table angle} = \frac{3.141592654 \times 4.5 \, (\text{circumference})}{16 \, (\text{pitch})}$$

$$= 0.8835729338$$

71

Thus, the corresponding angle is approximately 41.46294384° (from the calculator's table of natural tangents).

If a protractor is used to measure the angle it will be 41°27.8' or 41°28'.

Lead of the Machine

To cut a helix or spiral, the table feed screw is connected to the spindle through a train of change gears; therefore, for a given gear combination, the table advances a definite distance during each complete revolution of the spindle of the index head. If the change gears (which can be compared to those of the lathe) are all the same size, so that they do not change the velocity ratio between the table feed screw and the index head spindle, the table travel (in inches) per revolution of the index head spindle is the *lead of the machine*, which is identical to the *pitch of the machine*. The selection of the correct combination of change gears is important.

Thus, if the velocity ratio between the table feed screw and the index head spindle is unchanged, the cutter will mill a helix that has a pitch equal to the lead of the machine. If it is desirable to mill a helix that has a pitch different from the lead of the machine, change gears can be interposed to change the velocity ratio, so that a helix of the desired pitch can be produced.

If the lead, or pitch, of the table feed screw is ¼ inch (4 threads per inch), the worm rotates 40 turns to one turn of the worm wheel, which is attached to the index head spindle. Thus, the change gears all have the same diameter, and the lead of the machine is the standard 10 inches (40 × ¼). This means that the table advances a distance of ¼ inch per revolution of the table feed screw. As the table feed screw makes 40 revolutions to one revolution of the index head spindle, the lead of the machine is 10 inches.

To mill a helix that has a pitch less than 10 inches, change gears that increase the speed of the worm shaft must be interposed—or decrease the speed of the worm shaft for a pitch greater than 10 inches.

CHANGE-GEAR CALCULATIONS

The corresponding velocity ratios must be calculated for the different pitches that can be used to mill the various kinds of helix.

To meet these requirements the change gears can be arranged as: (1) simple gearing and (2) compound gearing.

Change Gears

For simple gearing, it is necessary only to select change gears that change the velocity ratio as:

$$\text{Velocity ratio} = \frac{\text{pitch of helix}}{\text{lead of machine}}$$

The change gears can be selected most conveniently by raising both terms of the ratio to correspond with the number of teeth of the change gears available. When the change gears cannot be selected in this manner, compound gearing must be used.

Change-gear train—The change-speed gear train is composed of four gears as follows: (1) the gear on the table feed screw shaft; (2) the first stud gear, so called because it is the first gear to be positioned; (3) the second stud gear; and (4) the gear on the worm. The gear on the worm is somewhat a misnomer, as it is not actually the gear on the worm. It is the gear on a shaft that has a bevel gear on the opposite end of the shaft that meshes with another bevel gear of the same size on the worm shaft. For lack of a better name and because there is no change in the velocity ratio, the result is equivalent to its being placed directly on the worm shaft.

The four gears in the gear train are illustrated in Figs. 3-6 and 3-7. The gear on the table screw shaft and the first stud gear are the *driver* gears; the second stud gear and the gear on the worm are the *driven* gears.

Change-gear ratio—Different combinations of change gears can be used to determine the distance that the table moves during one revolution of the spindle. The pitch of the helix to be milled depends on the change-gear ratio. Expressed as a formula, the change-gear ratio can be determined as follows:

$$\text{change-gear ratio} = \frac{\text{pitch of helix}}{\text{lead of the machine}}$$

Example: If the lead of the machine is 10 inches, what change-gear ratio is required to cut a helix that has a pitch of 10.50 inches.? Substituting in the formula:

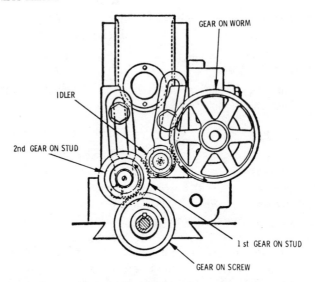

Fig. 3-6. Diagram of change gearing, showing the use of the idler.

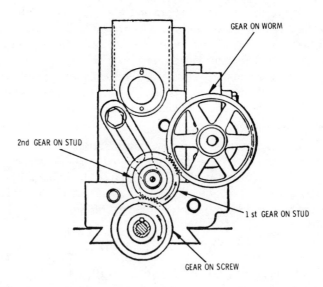

Fig. 3-7. Change gearing that requires no idler.

74

$$\text{Change-gear ratio} = \frac{10.5}{10}$$

Multiply by 10 to get

$$\frac{105}{100}$$

Then divide by 5, top and bottom, to get

$$\frac{21}{20}$$

or multiply this by 2 to produce

$$\frac{42}{40}$$

Therefore, a 20-tooth gear is placed on the table feed screw, and a 21-tooth gear is placed on the worm-shaft extension. If connected by idlers, the combination would provide the 10.5 pitch for the helix. However, there are no gears on the list of available gears that have 20 or 21 teeth. A 40-tooth gear is available in the set, but a 42-tooth gear is not available. Thus, calculations for equivalent gears must be made.

Change-gear calculations—Basically, these calculations are the same as for the change gears of an engine lathe. If change gears having the same diameter are used, a helix that has a pirch equal to the lead of the machine (standard lead is 10 inches) will be produced.

Equations illustrating the relationship of the different values are:

$$\text{Change-gear ratio} = \frac{\text{pitch of helix}}{\text{lead of machine}}$$

$$= \frac{\text{pitch of helix}}{10 \text{ (standard lead)}}$$

and:

$$\frac{\text{Driven gears}}{\text{Driver gears}} = \frac{\text{pitch of helix}}{10}$$

75

Since the product of each kind of gears determines the change-gear ratio:

$$\frac{\text{Product of driven gears}}{\text{Product of driver gears}} = \frac{\text{pitch of required helix}}{10}$$

The compound ratio of the driven gears can always be represented by a fraction. The numerator indicates the pitch to be cut, and the denominator indicates the lead of the machine. For example, if the required pitch is 20 inches and the lead of the machine is 10 inches (standard), the ratio is 20:10. Expressed in units, the ratio is the same as one-tenth of the required pitch to one. A convenient means of remembering the ratio is as follows: If the pitch is 40, the ratio of the gears is 4:1; if the pitch is 25, the ratio is 2.5:1; etc.

Example: Determine the necessary gears to be used in milling a helix that requires a 12-inch pitch.

The compound ratio of the driven to the driver gears is:

$$\frac{\text{Product of driven gears}}{\text{Product of driver gears}} = \frac{\text{pitch of required helix}}{10} = \frac{12}{10}$$

This fraction can be resolved into factors to represent the two kinds of change gears as follows:

$$\frac{12}{10} = \frac{(3 \times 4)}{(2 \times 5)}$$

Then, each term can be multiplied by a number common to both—24, in this instance—so that the numerator and denominator will correspond to the number of teeth of two change gears that are available with the machine. These multiplications do not affect the value of the fraction as:

$$\frac{3 \times 24}{2 \times 24} = \frac{72}{48}$$

Likewise, the second pair of factors can be treated similarly:

$$\frac{4 \times 8}{5 \times 8} = \frac{32}{40}$$

Therefore, the driven gears (72 and 32) and the driver gears (48 and 40) are selected as follows:

$$\frac{12}{10} = \frac{\text{product of driven gears}}{\text{product of driver gears}} = \frac{72 \times 32}{48 \times 40}$$

As has been indicated, the first selected pair of gears (72 and 32) are the driven gears because the numerators of the fractions represent the driven gears (gear on worm and the second stud gear); therefore, the 72-tooth gear is placed on the worm, and the 32-tooth gear is the second stud gear. The second pair of gears (48 and 40) are the driver gears because the denominators of the fractions represent the driver gears (gear on table feed screw and the first stud gear); therefore, the 48-tooth gear is placed on the table feed shaft, and the 40-tooth gear is the first stud gear.

The steps for determining the change gears required to cut a helix having a given pitch can be summarized as follows:

1. Determine the ratio between the required pitch of the helix and the lead of the machine (10 is a standard lead).
2. Express the ratio in the form of a fraction.
3. Resolve the fraction into two factors.
4. Raise the factors to higher terms, so that they correspond to the number of teeth in gears that are available with the machine.
5. The numerators represent the driven gears (gear on worm and second stud gear).
6. The denominators represent the driver gears (gear on feed screw and first stud gear).
7. Add an idler gear to cut a left-hand helix (on most machines).

Example: Select the gears for cutting a helix with a pitch of 27 inches.

$$\frac{27}{10} = \frac{3}{2} \times \frac{9}{5} = \frac{3}{2} \times \frac{16}{16} \times \frac{9}{5} \times \frac{8}{8} = \frac{48 \times 72}{32 \times 40}$$

The gear on the worm and the second stud gear are the 48-tooth gear and the 72-tooth gear, respectively. The gear on the table feed screw and the first stud gear are the 32-tooth gear and the 40-tooth gear, respectively.

Change-gear calculations can be checked by multiplying the product of the driven gears (48 $\times$ 72) by 10, and dividing by the product of the driver gears (32 × 40). The quotient is equal to the pitch of the resulting helix:

$$\frac{48 \times 72}{32 \times 40} \times 10 \ = \ 27 \text{ inches pitch}$$

The above check is derived from the fact that the quotient of the product of the driven gears divided by the product of the driver gears is equal to the pitch of the helix divided by 10 (standard lead of the machine), or one-tenth of the pitch. Thus, ten times the product of the driven gears divided by the product of the driver gears is equal to the pitch of the helix.

MILLING A SPIRAL

When tapered reamers, bevel gears, etc., are to be held between centers and milled, that is, the cuts are to be taken at an angle to the axis of the work, the axis of the index head and the tailstock center should coincide with the axis of the work. If they do not coincide, errors in indexing and problems in machining are introduced. A typical setup for milling tapered work is shown in Fig. 3-8.

Either a tilting table, an adjustable tailstock, or a taper attachment can be used to mill a piece of work that is tapered. These devices aid in mounting the work correctly (Fig. 3-9).

The taper attachment (see Fig. 3-9) has one end attached to the spindle; the opposite end is bolted to a slotted bracket that is mounted on the table as shown in the diagram. If neither the tilting table nor the taper attachment is available, several objectionable methods of mounting the tapered workpiece are often employed. Sometimes the tailstock is blocked up to the required height—with the index head having no angular adjustment (Fig. 3-10). In this arrangement the work does not bear properly on the centers, and errors are introduced because of the angularity of the dog and the reciprocating motion of the tail of the dog in the slot of the driver. Misalignment of centers results in an uneven and wobbly bearing.

In milling machine work, there should be no lost motion between the tail of the dog and the driver plate. However, as shown in Fig. 3-10, it is necessary to clamp the tailstock loosely to

Courtesy Cincinnati Milacron Co.

Fig. 3-8. Setup for milling tapered work on the milling machine.

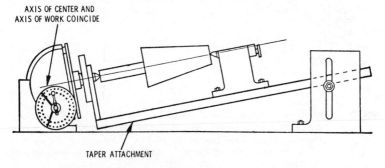

Fig. 3-9. Using the taper attachment to mill tapered work.

allow for the reciprocating motion of the tail of the dog, which is caused by the angularity of the dog. The angularity of the dog causes variation in the angular motion of the spindle and the work; thus, indexing errors are introduced.

To index the work at an angle of 180° (*A* and *B* in Fig. 3-11), the spindle would have to be indexed either more or less than 180° (*C*

79

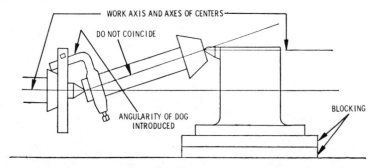

Fig. 3-10. Using blocks to raise the tailstock for milling tapered work. This practice is objectionable but is used in the absence of a tilting table or taper attachment.

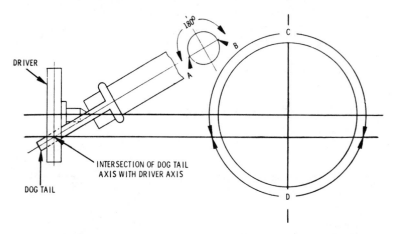

Fig. 3-11. Note the variation of the rotation of the work. This is due to the angularity of the work and is caused by the angularity of the tail of the dog. If the head is indexed at 180°, rotation of the work is either more or less than 180°, depending on the direction of rotation.

and D in Fig. 3-11), depending on the direction of rotation. This is because of the angularity of the dog.

SUMMARY

A spiral is a curve generated from a point that has three distinctive motions. These three distinctive motions are rotation about

80

the axis, advancement parallel with the axis, and increasing or decreasing distance from the axis or radius. A helix is a curve generated from a point that both rotates and advances axially on a cylindrical surface. The lead screw on a lathe is an example of a helix.

The pitch, or lead, of a helix is the distance the table travels as the work is rotated by the index head through a complete revolution. The two terms are identical in meaning. Pitch is probably a more proper term, but lead is more commonly used in the machine shop.

When milling a helix, the table angle is equal to the angle of the helix. This angle is the angle which the table must be turned to cut a helix. Two methods can be used to determine the table angle for cutting a helix.

REVIEW QUESTIONS

1. Explain the term helix.
2. Explain the term spiral.
3. What is the difference between pitch and lead? Explain.
4. What is the change-gear train?
5. What effect does an idler gear have when placed in the change-gear train?

CHAPTER 4

Spur Gear Computations

A gear is a form of disk, or wheel, having teeth around its periphery for providing a positive drive by meshing these teeth with similar teeth on another gear or rack. The slipping action of a belt, or other drive depending on friction, is eliminated with such a gear arrangement. A small amount of lost motion or *backlash* occurs between any two connected gears, but this is taken up by movement of the driving gear to give positive drive. In any two connected gears, the gear that receives the power is the *driver gear*, and the gear to which the power is delivered is the *driven gear*.

EVOLUTION OF GEARS

A smooth cylinder mounted on a shaft may be considered to be a gear having an infinite number of teeth. The fundamental principal of toothed gearing is illustrated by a pair of cylinders mounted on parallel shafts with their surfaces in contact and rolling together in opposite directions (Fig. 4-1).

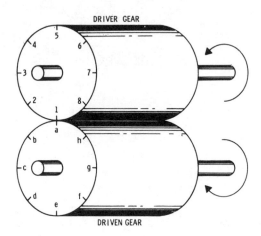

Fig. 4-1. The evolution of gears. Two cylinders in rolling contact.

As the teeth on a smooth cylinder are infinitely small, they do not project above the cylindrical surface. If power is applied to the driver cylinder in the direction of the arrow, the driven cylinder will turn by friction. This friction is equivalent to the meshing of infinitely small teeth. If the cylinders are of equal size, they will turn at the same speed—that is, the equally spaced divisions (1, 2, 3, etc.) will coincide as the two cylinders rotate. This is true only as long as the load on the driven cylinder is not large enough to cause slippage. In this particular instance, very little friction is present to prevent slippage of the cylinders (line *EL* in Fig. 4-2).

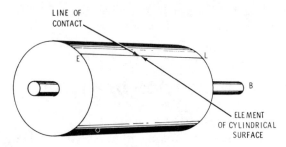

Fig. 4-2. The driven gear, showing the line of contact *EL*.

If the cylindrical surfaces were perfectly smooth, frictional contact would not be present to turn the cylinders. Although metal surfaces may appear smooth to the eye and to touch, minute irregularities are present even though they are invisible without magnification. Thus, the line of contact (line *EL* in Fig. 4-3) has the

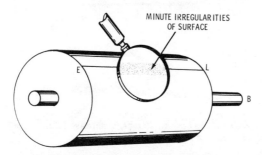

MINUTE IRREGULARITIES
OF SURFACE

Fig. 4-3. Note the minute irregularities of the surface in line of contact *EL*.

appearance of a strip of coarse emery paper. When the two cylinders are in contact, the interlocking minute irregularities produce the frictional contact. When pressure is applied to force a firm contact, friction is increased by the flexing or flattening of the metal along the line of contact, thus increasing the contact area (Fig. 4-4). This exaggerated illustration of flexing of surfaces may be compared to the action of a clothes wringer with its two rubber rolls under considerable pressure.

In the perfect surface, there are no minute irregularities or flexing; therefore, there is no frictional contact, and the line of contact (line *EL* in Fig. 4-3) is a part of the surface—that is, it has only one dimension (length), with no contact area. If machined surfaces were perfect surfaces, frictional contact would be impossible, and power could not be transmitted by cylinders. Hence, the necessity for toothed gears to obtain a positive drive is evident.

GEAR TEETH

The position of the teeth with respect to the periphery of the cylinders should be understood before considering the various shapes of gear teeth and the method of generating these shapes. It

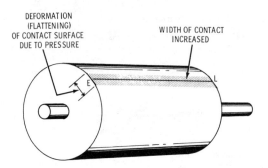

DEFORMATION
(FLATTENING)
OF CONTACT SURFACE
DUE TO PRESSURE

WIDTH OF CONTACT
INCREASED

E

L

Fig. 4-4. The width of the contact line *EL* is increased due to the flattening of the surface under pressure.

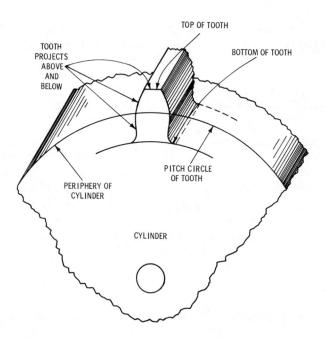

TOP OF TOOTH

BOTTOM OF TOOTH

TOOTH
PROJECTS
ABOVE
AND
BELOW

PITCH CIRCLE
OF TOOTH

PERIPHERY OF
CYLINDER

CYLINDER

Fig. 4-5. Position of gear tooth with respect to the periphery of the cylinder.

86

should be noted that each tooth projects both above and below the periphery of the cylinder (Fig. 4-5). If the surfaces of two cylinders are to remain in contact, teeth could not be formed in their surfaces by cutting grooves in them because it would be necessary to move their axes closer together for the teeth to interlock; thus the original surfaces would overlap. Teeth could not be added to the cylinders because the axes would have to be moved farther apart, thus separating the contact surfaces.

Therefore, a combination of the two methods must be used: cutting grooves equal to one-half the proposed depth plus clearance, and adding an equal amount between the spaces formed to complete the partly formed teeth. Then the teeth will fall into the spaces and interlock properly, the original surfaces of the cylinders will remain on the contact line, and the diameters of the cylinders will provide the main circles for all calculations for speed, numbers, teeth dimensions, etc.

Gear Tooth Terms

Of course, if the cylinders were gear blanks on which gear teeth were to be cut, the teeth could not project above the cylindrical surfaces of the blanks, but this is done in Fig. 4-5 for purposes of illustration.

Pitch circle—As mentioned, the original surfaces of the cylinders remain on the contact line, and the diameters of these cylinders, or circles, provide the basis for the various gear tooth computations. The pitch circle is the line of contact of the two cylinders. The pitch circle is the reference circle of measurement, and is located one-half the distance between the top and bottom of the theoretical tooth (Fig. 4-6).

The *pitch point* is the point of tangency of two pitch circles—or of a pitch circle and a pitch line—and is located on the line of centers. The point of intersection of a tooth profile with the pitch circle is its pitch point.

The *face* of a gear tooth is the surface of the tooth between the pitch circle and the top of the tooth. The *flank* of the tooth is the surface of the tooth between the pitch circle and the bottom of the groove, including the *fillet*. The fillet is a small arc or *fillet curve* that joins the tooth profile to the bottom of the tooth space, thus

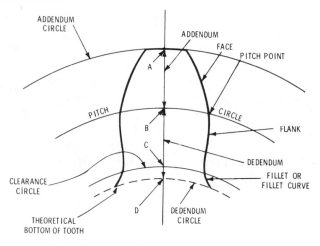

Fig. 4-6. Diagram of a theoretical gear tooth.

avoiding sharp corners at the root of the tooth.

Addendum circle and dedendum circle—The addendum circle is the circle that passes through the top of the gear teeth; the diameter of the addendum circle is the same as the *outside diameter* of the gear. The *addendum* is the height of the tooth above the pitch circle, or the radial distance between the pitch circle and the top of the tooth. The circle that passes through the bottom of the tooth space is called the *dedendum circle*. The *dedendum* is the depth of the tooth space below the pitch circle, or the radial dimension between the pitch circle and the bottom of the tooth space (see Fig. 4-6).

Spur Gear Computations

The spur gear is the simplest gear and the one in most common use. A spur gear has straight teeth cut parallel with the axis rotation of the gear body. All the other gear forms—bevel gears, helical gears, worm gears, and worm wheels (Fig. 4-7)—are modifications of the spur gear. The general principle, or the principle on which gear teeth are formed, is practically the same in all the forms of gears in use.

Circular pitch—The circular pitch is the distance from the center of one tooth to the center of the adjacent tooth as measured on

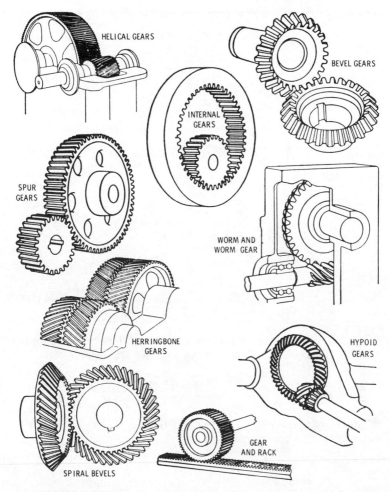

HELICAL GEARS

BEVEL GEARS

INTERNAL GEARS

SPUR GEARS

WORM AND WORM GEAR

HERRINGBONE GEARS

HYPOID GEARS

GEAR AND RACK

SPIRAL BEVELS

Fig. 4-7. Type of gears.

the pitch circle. It may be considered as the width of one tooth plus the width of one space as measured on the pitch circle. Thus, the circular pitch is an arc whose length depends on the number of teeth in the gear and on the diameter of the pitch circle. In Fig. 4-8 it may be noted that the circular pitch is equal to the length of the arc ABC. This arc is equal in length to the length of the arc EF, also on the pitch circle. Circular pitch may be determined by dividing

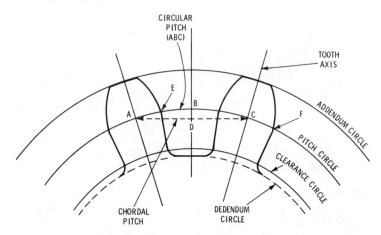

Fig. 4-8. Note the circular pitch and chordal pitch.

the circumference of the pitch circle (πD) by the number of teeth as follows:

$$\text{Circular pitch} = \frac{\text{circumference of pitch circle}}{\text{number of teeth}}$$

If the circumference of the pitch circle and the circular pitch are known, the number of teeth in the gear may be calculated:

$$\text{Number of teeth} = \frac{\text{circumference of pitch circle}}{\text{circular pitch}}$$

Likewise, if the circular pitch and the number of teeth in the gear are known, the diameter of the pitch circle may be calculated as:

$$\text{Diameter of pitch circle} = \frac{\text{circular pitch} \times \text{number of teeth}}{3.1416}$$

Chordal pitch is the distance from the center of one tooth to the center of another tooth when measured on the chord of an arc of the pitch circle (Fig. 4-8).

The formulas for spur gears have been assembled together in Table 4-1 for convenience.

Diametral pitch—The number of teeth in a gear per inch of pitch circle diameter is called the diametral pitch. The diametral

Table 4-1. Formula for Spur Gear Calculations

No. of Teeth	Chordal Thickness	Chordal Addend.	No. of Teeth	Chordal Thickness	Chordal Addend.	No. of Teeth	Chordal Thickness	Chordal Addend.
10	1.56435	1.06156	59	1.57061	1.01046	108	1.57074	1.00570
11	1.56546	1.05598	60	1.57062	1.01029	109	1.57075	1.00565
12	1.56631	1.05133	61	1.57062	1.01011	110	1.57075	1.00560
13	1.56698	1.04739	62	1.57063	1.00994	111	1.57075	1.00556
14	1.56752	1.04401	63	1.57063	1.00978	112	1.57075	1.00551
15	1.56794	1.04109	64	1.57064	1.00963	113	1.57075	1.00546
16	1.56827	1.03852	65	1.57064	1.00947	114	1.57075	1.00541
17	1.56856	1.03625	66	1.57065	1.00933	115	1.57075	1.00537
18	1.56880	1.03425	67	1.57065	1.00920	116	1.57075	1.00533
19	1.56899	1.03244	68	1.57066	1.00907	117	1.57075	1.00529
20	1.56918	1.03083	69	1.57066	1.00893	118	1.57075	1.00524
21	1.56933	1.02936	70	1.57067	1.00880	119	1.57075	1.00519
22	1.56948	1.02803	71	1.57067	1.00867	120	1.57075	1.00515
23	1.56956	1.02681	72	1.57067	1.00855	121	1.57075	1.00511
24	1.56967	1.02569	73	1.57068	1.00843	122	1.57075	1.00507
25	1.56977	1.02466	74	1.57068	1.00832	123	1.57076	1.00503
26	1.56986	1.02371	75	1.57068	1.00821	124	1.57076	1.00499
27	1.56991	1.02284	76	1.57069	1.00810	125	1.57076	1.00495
28	1.56998	1.02202	77	1.57069	1.00799	126	1.57076	1.00491
29	1.57003	1.02127	78	1.57069	1.00789	127	1.57076	1.00487
30	1.57008	1.02055	79	1.57069	1.00780	128	1.57076	1.00483
31	1.57012	1.01990	80	1.57070	1.00772	129	1.57076	1.00479
32	1.57016	1.01926	81	1.57070	1.00762	130	1.57076	1.00475
33	1.57019	1.01869	82	1.57070	1.00752	131	1.57076	1.00472
34	1.57021	1.01813	83	1.57070	1.00743	132	1.57076	1.00469
35	1.57025	1.01762	84	1.57071	1.00734	133	1.57076	1.00466
36	1.57028	1.01714	85	1.57071	1.00725	134	1.57076	1.00462
37	1.57032	1.01667	86	1.57071	1.00716	135	1.57076	1.00457
38	1.57035	1.01623	87	1.57071	1.00708	136	1.57076	1.00454
39	1.57037	1.01582	88	1.57071	1.00700	137	1.57076	1.00451
40	1.57039	1.01542	89	1.57072	1.00693	138	1.57076	1.00447
41	1.57041	1.01504	90	1.57072	1.00686	139	1.57076	1.00444
42	1.57043	1.01471	91	1.57072	1.00679	140	1.57076	1.00441
43	1.57045	1.01434	92	1.57072	1.00672	141	1.57076	1.00439
44	1.57047	1.01404	93	1.57072	1.00665	142	1.57076	1.00435
45	1.57048	1.01370	94	1.57072	1.00658	143	1.57076	1.00432
46	1.57050	1.01341	95	1.57073	1.00651	144	1.57076	1.00429
47	1.57051	1.01311	96	1.57073	1.00644	145	1.57077	1.00425
48	1.57052	1.01285	97	1.57073	1.00637	146	1.57077	1.00422
49	1.57053	1.01258	98	1.57073	1.00630	147	1.57077	1.00419
50	1.57054	1.01233	99	1.57073	1.00623	148	1.57077	1.00416
51	1.57055	1.01209	100	1.57073	1.00617	149	1.57077	1.00413
52	1.57056	1.01187	101	1.57074	1.00611	150	1.57077	1.00411
53	1.57057	1.01165	102	1.57074	1.00605	151	1.57077	1.00409
54	1.57058	1.01143	103	1.57074	1.00599	152	1.57077	1.00407
55	1.57058	1.01121	104	1.57074	1.00593	153	1.57077	1.00405
56	1.57059	1.01102	105	1.57074	1.00587	154	1.57077	1.00402
57	1.57060	1.01083	106	1.57074	1.00581	155	1.57077	1.00400
58	1.57061	1.01064	107	1.57074	1.00575	156	1.57077	1.00397

pitch is a ratio of the number of teeth in a gear to the number of inches in the diameter of the pitch circle. As it is a ratio between two quantities, diametral pitch cannot be shown on a blueprint as a dimension. The diametral pitch system is designed to designate a series of gear tooth sizes (whole numbers), just as screw-thread pitches are standardized to designate the size of the thread. So, it should be remembered that there must be a whole number of teeth in a gear. Diametral pitch is usually referred to as *pitch*. For example, a 10-pitch gear indicates that a gear has 10 teeth per inch of diameter of its pitch circle.

The diametral pitch of a gear may be determined by dividing the number of teeth in the gear by its pitch circle diameter:

$$\text{Diametral pitch} = \frac{\text{number of teeth}}{\text{pitch circle diameter}}$$

Example: If a gear has 22 teeth, and the diameter of the pitch circle is 2 inches, calculate the diametral pitch.

$$\text{Diametral pitch} = \frac{22}{2} = 11$$

The following relationships may be obtained from the equation:

1. The number of teeth in the gear may be obtained by multiplying the diametral pitch by the diameter of the pitch circle as:

Number of teeth = diametral pitch $\times$ diameter of pitch circle

   Example: Given: diametral pitch = 11; diameter of pitch circle = 2. Find the number of teeth.

$$\text{Number of teeth} = 11 \times 2 = 22 \text{ teeth}$$

2. The diameter of the pitch circle may be obtained by dividing the number of teeth by the diametral pitch as:

$$\text{Diameter of pitch circle} = \frac{\text{number of teeth}}{\text{Diametral pitch}}$$

   Example: Given: number of teeth = 22; diametral pitch = 11. Find the diameter of the pitch circle.

$$\text{Diameter of pitch circle} = \frac{22}{11} = 2 \text{ inches}$$

Nearly all gear calculations are made in terms of diametral pitch rather than circular pitch because diametral pitch may usually be expressed as a whole number, which is convenient for expressing the proportions of teeth. A series of symbols and formulas may be used to determine the proportions of gear teeth (Fig. 4-9).

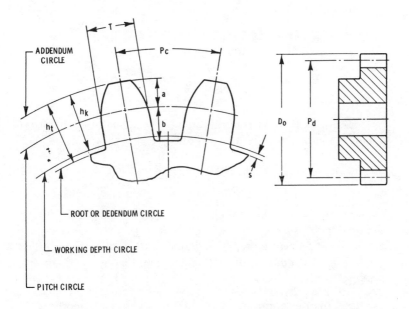

Courtesy Cincinnati Milacron Co.

Fig. 4-9. Diagram showing the proportions of spur gear teeth. These symbols are used in formulas for determining the proportion of spur gear teeth.

Frequently, it is convenient to change diametral pitch (P) to circular pitch (Pc), or vice versa. If either of the two terms is known, the other may be determined, as shown by the following rules and formulas:

1. To calculate diametral pitch (P) when circular pitch is shown, divide 3.1416 by the circular pitch (Pc):

$$P = \frac{3.1416}{Pc}$$

Example: If the circular pitch (Pc) of a gear is ⅝ (0.625) inch, what is the diametral pitch (P) to the nearest whole number?

$$P = \frac{3.1416}{0.625}$$

$$= 3.1416 \div 0.625$$

$$= 5.026 \text{ or } 5$$

2. To calculate circular pitch (Pc) when diametral pitch is known, divide 3.1416 by the diametral pitch (P):

$$Pc = \frac{3.1416}{P}$$

Example: If the diametral pitch (P) of a gear is 5, what is the corresponding circular pitch (Pc)?

$$Pc = \frac{3.1416}{5} = 0.6283 \text{ or } ⅝ \text{ inch}$$

Pitch diameter—The diameter of the pitch circle is called pitch diameter (Pd). The operating pitch diameter is the pitch diameter at which the gears operate. Pitch diameter (Pd) can be determined by dividing the number of teeth in the gear (N) by the diametral pitch (P) as:

P = diametral pitch	b = dedendum
Pc = circular pitch	h_k = working depth
Pd = pitch diameter	h_t = whole depth
Do = outside diameter	S = clearance
N = number of teeth in gear	C = center distance
T = tooth thickness	L = length of rack
a = addendum	

$$Pd = \frac{N}{P}$$

Example: If the gear has 22 teeth (N) and the diametral pitch is 11, what is the pitch diameter (Pd)?

$$Pd = \frac{22}{11} = 2 \text{ inches}$$

Addendum—The height by which a tooth projects above the pitch circle is called the addendum. It is also the radial distance between the pitch circle and the addendum circle (see Fig. 4-9). In spur gear calculations, the addendum (a) is equal to 1.0 divided by the diametral pitch (P):

$$a = \frac{1.0}{P}$$

Example: If the diametral pitch (P) of a gear is 11, what is the addendum (a), or height above the pitch circle?

$$a = \frac{1.0}{11} = 0.0909 \text{ inch}$$

Dedendum—The depth of a tooth space below the pitch circle is called the dedendum; or it is the radial distance between the pitch circle and the root or dedendum circle (see Fig. 4-8). The dedendum (b) is calculated by dividing 1.157 by the diametral pitch (P):

$$b = \frac{1.157}{P}$$

Example: Calculate the dedendum (b) for a gear having a diametral pitch (P) of 11.

$$b = \frac{1.157}{11} = 0.1052 \text{ inch}$$

Number of teeth in a gear—As the diametral pitch expresses the number of teeth per inch of pitch circle diameter, it follows that the number of teeth in a gear (N) can be found by multiplying the diameter of the pitch circle (Pd) by the diametral pitch (P):

$$N = PdP$$

Example: If the diametral pitch (P) of a gear is 10 and the pitch circle diameter (Pd) is 2 inches, how many teeth (N) are in the gear?

$$N = 2 \times 10 = 20 \text{ teeth}$$

Likewise, the diameter of the pitch circle (Pd) may be found by dividing the number of teeth (N) in the gear by the diametral pitch (P):

$$Pd = \frac{N}{P}$$

Example: If there are 20 teeth in a gear (N), and the diametral pitch (P) is 10, what is the pitch diameter (Pd)?

$$Pd = \frac{20}{10} = 2 \text{ inches}$$

Clearance—As mentioned previously, gears must be so proportioned that when a tooth is in mesh with the mating gear, the top of the tooth should not touch the bottom of the groove or space between the tooth and the adjacent tooth. The clearance, or radial distance from the dedendum circle to the clearance circle (see Fig. 4-9), provides a margin of space to allow for any slight errors in machining. The clearance is the amount by which the dedendum exceeds the addendum of a given gear. In actual practice, the clearance is equal to one-twentieth ($\frac{1}{20}$) of the circular pitch. In gear calculations, clearance (S) may be determined by dividing 0.157 (obtained by taking $\frac{1}{20}$ of 3.1416) by the diametral pitch (P) as shown by the formula:

$$S = \frac{0.157}{P}$$

Example: Calculate the clearance (S) for a gear having a diametral pitch (P) of 11.

$$S = \frac{0.157}{11} = 0.0143 \text{ inch clearance}$$

Working depth—The theoretical length of the tooth is the working depth. The working depth of a tooth is equal to twice the addendum (see Fig. 4-9). For two mating gears, the working depth is the sum of their addenda or the depth of engagement of the two gears. The working depth (h_k) of a gear may be calculated by dividing 2.0 by the diametral pitch (P) as shown by the formula:

$$h_k = \frac{2.0}{P}$$

Example: If the diametral pitch (P) of a gear is 11, determine the working depth (h_k).

$$h_k = \frac{2.0}{11} = 0.1818 \text{ inch}$$

Whole depth—This is the total depth of a tooth space and is equal to the addendum plus the dedendum. The whole depth is also equal to the working depth plus clearance (see Fig. 4-9). The whole depth (h_t) can be determined by dividing 2.157 by the diametral pitch (P):

$$h_t = \frac{2.157}{P}$$

Example: If the diametral pitch (P) of a gear is 11, find the whole depth (h_t).

$$h_t = \frac{2.157}{11} = 0.196 \text{ inch}$$

Outside diameter—The diameter of an addendum circle (see Fig. 4-9) is the outside diameter of a gear. As the height of the teeth above the pitch circle is equal to 1 divided by the diametral pitch, the formula for outside diameter can be expressed as:

$$D_o = Pd + \frac{1}{P} + \frac{1}{P}$$

A 20-tooth, 10-pitch gear would have a height above the pitch circle equal to $\frac{1}{10}$, or 0.1 inch (Fig. 4-10). The diameter of the pitch circle is equal to $\frac{20}{10}$, or 2 inches. Thus, the outside diameter of the gear is equal to 0.1 + 0.1 + 2.0 = 2.2 inches.

The outside diameter of the gear (2.2 inches) is equal to the pitch diameter of a gear with two additional teeth (22 divided by 10 = $\frac{22}{10}$ = 2.2 inches). Therefore, the outside diameter (D_o) of a gear may be calculated by adding 2 to the number of teeth (N) and dividing the sum by the diametral pitch (P):

$$D_o = \frac{N}{P} + \frac{1}{P} + \frac{1}{P}$$

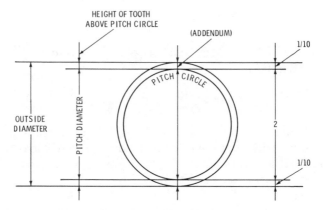

Fig. 4-10. Diagram showing outside diameter of gear teeth.

$$Do = \frac{N + 2}{P}$$

Example: If a gear has 20 teeth (N) and the diametral pitch (P) is 10, find the outside diameter (Do) of the tooth.

$$Do = \frac{20 = 2}{10} = \frac{22}{10} = 2.2 \text{ inches}$$

Tooth thickness—This is the length of an arc of the pitch circle between the two sides of a gear tooth. Tooth thickness (T) may be calculated by dividing 1.5708 by the diametral pitch (P):

$$T = \frac{1.5708}{P}$$

Example: Find the tooth thickness (T) if the diametral pitch (P) is 11.

$$T = \frac{1.5708}{11} = 0.1428 \text{ inch}$$

Side clearance (backlash)—Theoretically, the thickness of a tooth on the pitch circle is the same as the width of the groove. In practice, it is necessary to make the width of the space slightly larger than the thickness of the tooth to allow for inaccuracies of operation and workmanship. The difference in the thickness of the tooth and the width of the space is called backlash.

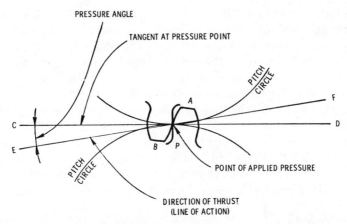

Fig. 4-11. Diagram showing the pressure angle.

Center distance—As the diametral pitch designates the number of teeth in a gear per inch of diameter, the center distance (C) between two meshing gears can be determined by adding the number of teeth in both gears (N_1 + N_2) and dividing the sum by two times the diametral pitch (P):

$$C = \frac{(N_1 + N_2)}{2P}$$

Example: If the diametral pitch (P) is 4, one gear has 60 teeth (N_1), and the other gear has 20 teeth (N_2), what is the distance between centers (C)?

$$C = \frac{60 + 20}{2(4)} = \frac{80}{8} = 10 \text{ inches}$$

Length of rack—A rack is a gear with its teeth spaced along a straight line, making it suitable for straight-line motion. The length of a rack (L) may be determined by multiplying the number of teeth in the rack (N) by the circular pitch (Pc): L = NPc.

Pressure angle—This is the angle between the tooth profile and a radial line at its pitch point. In involute teeth, the pressure angle is described as the angle between the line of action and the line tangent to the pitch circle (Fig. 4-11).

In Fig. 4-11, the pitch circles of the two gears make contact at the pitch point *P*. Two mating teeth are shown in contact at that point on the pitch circles. The line *CD* is tangent to the pitch circles through pitch point *P*, and the line *EF* represents the direction of the applied pressure of the driving tooth (line of action). Angle *EPC*, which is between the line of action *EF* and the tangent line *CD*, is called the pressure angle.

In gear design, a pressure angle of 14.5° is most commonly used for involute teeth. It has been found by experience that maximum efficiency can be obtained through a pressure angle of 14.5°. The selection of the 14.5° pressure angle was influenced, originally, by the fact that the sine of 14.5° is approximately 0.25, a convenient proportion for the millwright to lay out. The involute stub tooth has a standard pressure angle of 20°. (Actually, sine of 14.5° = 0.2503800041.)

In gear design, the pressure angles are either 14.5° or 20°. Formerly, it was believed that a pressure angle of 20° caused too much wear on the gear teeth and a rougher action between the gears. However, it has been found that the bearing pressure is not greater, wear on the teeth is no more, and the action is just as smooth for gears with a 20° pressure angle.

Involute Gears

The involute curve is used almost exclusively for gear tooth profiles. The form or shape of an involute curve is dependent on the diameter of the base circle from which it is derived.

An involute curve may be traced by a point on a taut string which is unwound from the circumference of a circle—the base circle of the involute. The generation of an involute curve is shown in Fig. 4-12.

To describe an involute of a given base circle let *A, B, C*, etc., be equal divisions on the circumference of the circle. At each of these points draw tangents to the circle. On the tangent at *B*, make a distance *Bb* equal to the arc *AB*; on the tangent *C*, make the distance *Cc* = 2 times arc *AB*, etc. Thus, the points *b, c, d*, etc., are obtained. Through these points describe the curve *A, b, c, d, e, f, g*, which is the involute of the base circle.

To illustrate the generation of the involute form of tooth (Fig. 4-13), first describe the pitch circle, addendum circle, and dedendum circle with radii equal to the selected data for the tooth to be

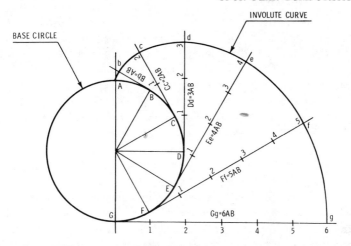

Fig. 4-12. Diagram showing the method of describing an involute curve from a given base circle.

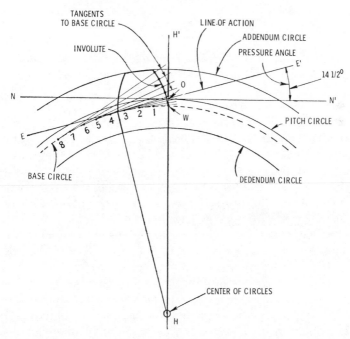

Fig. 4-13. Diagram showing the method of generating an involute gear tooth.

drawn. Select any point O as the pitch point on the pitch circle. Draw the line HH' from the center through the pitch point O, and also through point O draw line NN' at an angle of 90° to HH'. Line NN' is then tangent to the pitch circle at point O.

Through point O, draw the line of action EE', at an angle of $14\frac{1}{2}°$ with line NN' (angle $E'ON' = 14\frac{1}{2}°$). This is the pressure angle generally used. With H as the center, describe the base circle with a radius that will make the base circle tangent to the line of action EE'. To obtain points on the involute curve, draw tangents to the base circle at the equally spaced points $1, 2, 3, 4$, etc., and locate the points on the involute, as in Fig. 4-12.

It should be noted that the base circle is always smaller than the pitch circle. The involute curve forms the addendum of the tooth and extends to the base circle. The balance of the tooth, or flank, is drawn radially from the base circle, except for a small fillet at the bottom of the tooth.

SUMMARY

A gear is a form of disk, or wheel, having teeth around its periphery for providing a positive drive. In any two connecting gears, the gear that receives the power is the driver gear, and the gear to which the power is delivered is the driven gear.

The positions of gears with respect to the periphery of the cylinders should be understood before considering the various shapes of gear teeth and the method of generating these shapes. Each tooth projects both above and below the periphery of the cylinder. If the surfaces of two cylinders are to remain in contact, teeth could not be formed in their surfaces by cutting grooves in them.

The spur gear is the simplest gear and the one most commonly used. The spur gear has straight teeth cut parallel with the axis of rotation of the gear body. All other gear forms, such as bevel, helical, and worm, are modifications of the spur gear.

The number of teeth in a gear per inch of pitch circle diameter is called diametral pitch. The diametral pitch is a ratio of the number of teeth in a gear to the number of inches in the diameter of the pitch circle. The diametral pitch of a gear may be determined by dividing the number of teeth in the gear by its pitch circle diameter. Nearly all gear calculations are made in terms of diametral pitch rather than circular pitch.

The height by which a tooth projects above the pitch circle is called the addendum. It is also the radial distance between the pitch circle and the addendum circle. In spur gear calculations, the addendum is equal to 1.0 divided by the diametral pitch. The depth of a tooth space below the pitch circle is called the dedendum. It is the radial distance between the pitch circle and the root or dedendum circle. The dedendum is calculated by dividing 1.157 by the diametral pitch.

REVIEW QUESTIONS

1. What is the simplest form of gear?
2. What determines the number of teeth in a gear?
3. What is the addendum of a gear?
4. What is the dedendum of a gear?
5. What is the diametral pitch?

CHAPTER 5

Gears and Gear Cutting

The spur gear is the simplest gear and the one most commonly used. A spur gear has straight teeth that are cut parallel with the axis of rotation of the gear body. All the other gear forms—bevel gears, helical gears, worm gears, and worm wheels—are modifications of the spur gear. The general principle, or the principle on which gear teeth are formed, is practically the same in all the forms of gears in use.

DEVELOPMENT OF GEAR TEETH

The curves for gear tooth profiles should be constructed in such a way that a uniform velocity ratio is attained when two wheels are geared together. The common normal at the point of contact should always pass through the pitch point, which divides the line of centers in the inverse ratio of the angular velocities.

The shapes of gear teeth differ with the system used in generating them. The two systems in general use are: (1) involute and (2) cycloidal.

The *involute* curve is used almost exclusively for gear tooth profiles. The term involute refers to the shape of the curve. The form or shape of an involute curve is dependent on the diameter of the base circle from which it is derived. The curve can be traced by a point on a taut string which is unwound from a circumference of a circle—the base circle of the involute.

When two involute curves are brought into contact on the line of centers, a pitch point is established. The pitch point determines the diameter of the pitch circle. It is only in the involute form of tooth that the diameter of the pitch circle can be a flexible dimension, which permits involute gears to operate successfully at center distances that vary slightly.

Formerly, the *cycloidal* system was used almost exclusively. In the cycloidal system of generating gear teeth, the tooth profile is a double curve consisting of: (1) an epicycloidal face and (2) a hypocycloidal flank. The double curve is generated by rolling circles which roll on the pitch circle. The epicycloidal portion of the curve is generated by an outer rolling circle, and the hypocycloidal portion is generated by an inner rolling circle. The diameters of the rolling circles should not be greater than the radius of the pitch circle to avoid generating a gear tooth that is weak at the root. Also, the diameter of the rolling circle should not be less than one-half the radius of the pitch circle (Fig. 5-1).

If the diameter of the rolling circle is equal to the radius of the pitch circle, the hypocycloid will be a straight line that passes through the center of the pitch circle, and the flank of the tooth will be a radial. The same rolling circle must be used for interchangeable gears. In any other instance, the same rolling circles can be used for both gears, or their diameters can be used in proportion to their respective pitch circles.

A *cycloid* curve is described when the generating circle rolls along a straight line. An *epicycloid* curve is described when the generating circle (outer rolling circle) rolls along the outside of a circle. A *hypocycloid* curve is described when the generating circle (inner rolling circle) rolls along the inside of a circle.

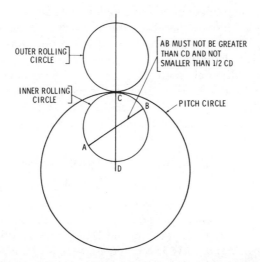

Fig. 5-1. Diagram showing the generation of cycloid form of gear teeth.

In comparing cycloid teeth with involute teeth, the involute gear teeth are stronger, run well with their centers at varying distances, and transmit uniform velocity. The chief objection to involute teeth is that they cause increased pressure on the bearings because of the obliquity of their action. However, modified tooth forms have been introduced on some involute gears.

Diametral and Circular Pitch Systems

Most of the cut gears in the United States are produced by the diametral pitch system. The circular pitch system is commonly used if the gear teeth are larger than *one* diametral pitch. Circular pitch can also be applied to smaller gears if the required center-to-center distance cannot be obtained by a standard diametral pitch. The circular pitch system is also used on cast gears and on worm gearing, although these gears can also be designed for diametral pitch.

As mentioned, the diametral pitch system can be used to designate a series of standard gear tooth sizes (whole numbers), just as screw-thread pitches are standardized to designate the size of a thread. As there must be a whole number of teeth on a gear, the increase in pitch diameter per tooth varies with the pitch. For example, the pitch diameter of a gear having *20* teeth of *4* diametral pitch is *5* inches; therefore, the increase in diameter for each additional tooth is equal to ¼ inch per tooth. Similarly, for *2* diametral pitch the increase in diameter would be equal to ½ inch for each additional tooth. If a given center distance must be maintained and a standard diametral pitch cannot be used, it may be necessary to use gears based on the circular pitch system. Gear teeth of different diametral pitch are shown in Fig. 5-2.

American Standard Spur Gear Tooth Forms

Four spur gear tooth forms are covered by the American Standards Association (ASA). As the rack is the basis of a standard system of interchangeable gears, it is necessary only to give the proportions of the rack teeth in establishing a gear tooth standard.

American Standard 14.5° Involute Full-Depth Tooth—The rack of a standard 14.5° full-depth standard tooth form is shown in Fig. 5-3. This tooth form is very successful if the tooth numbers are large enough to avoid excessive undercutting of the teeth. Undercutting begins when the number of teeth is less than 32 teeth.

American Standard 20° Involute Full-Depth Tooth—The formulas for the l4.5° and 20° full-depth tooth standards are identical except for the radius at the base of the tooth. The pressure angle is the chief difference in the two standards. The larger pressure angle reduces undercutting, which begins when the number of teeth is less than 18, and may be excessive when the number of teeth is less than 14. The 20° teeth are wider at the base, which makes them stronger than the 14.5° teeth (see Fig. 5.3).

The *American Standard 20° Involute Fine-Pitch Tooth* is a tooth form used for gears of 20 diametral pitch and finer. It is the same as the 20° full-depth tooth form except for a slight increase in whole depth. The additional depth and clearance are required because the wear on fine-pitch teeth is proportionally greater, the fillet radius is usually greater, and provision must be made for foreign material to accumulate at the bottom of the tooth spaces.

American Standard 14.5° Composite Tooth—This standard differs from the 14.5° involute full-depth system in regard to the form of the basic rack; the pressure angle and the various formulas for determining the tooth depth, addendum, etc., are the same.

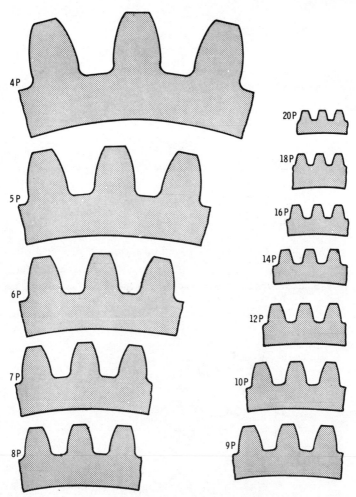

Courtesy Machinery's Handbook, Industrial Press

Fig. 5-2. Gear teeth of different diametral pitch (full size).

The involute form of rack is modified by introducing a cycloidal curve below the pitch line and one above it to make the tooth symmetrical, as required for interchangeable gearing.

As it is impractical to produce a rack with cycloidal curves in the shop, or a cutter of exact form, the approximate rack form (Fig.

109

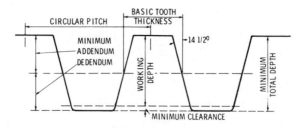

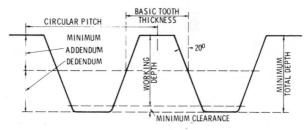

Fig. 5-3. Basic rack of the 14.5° full-depth involute system (top) and the 20°
full-depth involute system (bottom).

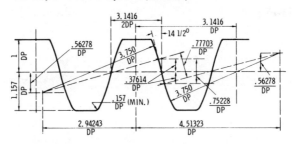

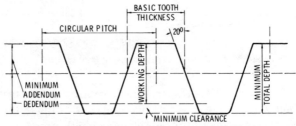

Fig. 5-4. Basic rack of the 14.5° composite system (top), and the 20° stub
involute system (bottom).

110

5-4) is used to meet the practical requirements. The curves are close approximations of the cycloidal curves on the theoretical rack.

The composite tooth form was developed originally for use with the form milling process; gear teeth conforming to the composite standard usually are cut by form milling. This form of teeth can be produced readily on hobbing machines and other machines by making a hob or cutter of the basic rack form. The relieving tool can be made to the form of the basic rack if a hob is used.

American Standard 20° Involute Stub Tooth—The chief difference between the stub tooth and the full-depth tooth form is in regard to the tooth depth (see Fig. 5-4). The shorter tooth, in combination with the 20° pressure angle, strengthens the stub form, and pinions with 12 and 13 teeth are undercut only slightly. The length of contact between mating gears is shortened, which tends to offset the increase in tooth strength and tends toward greater noise when the gears are running unless the tendency is offset by greater accuracy in cutting and mounting, This tendency to noise can be objectionable in certain classes of service, but it may be acceptable in other classes of service. For example, noise that is objectionable in automotive transmissions may not be a factor in gears used on some other classes of machinery.

The 20° stub tooth form is used extensively in automotive transmissions because the maximum power-transmitting capacity for a given pitch or material is essential and relatively small gears are required. However, very accurate gears are required, and the mountings are designed to minimize noise. Helical gear forms are also used because of their smooth continuous action. The American Gear Manufacturers' Association recommends the American Standard 20° stub tooth system. These gears can be used interchangeably with other stub tooth systems, and only the amount of clearance is affected by the result of variations in tooth heights.

Fellows Stub Tooth—Two diametral pitches are used as a basis for the system of stub gear teeth introduced by the Fellows Gear Shaper Co. One diametral pitch, say, 8, is used as the basis for obtaining the dimension for the addendum and dedendum; another diametral pitch, say, 6, is used to obtain the dimensions for the thickness of tooth, the number of teeth, and the pitch diameter. Thus, the teeth are designated as ⁶⁄₈ pitch; the numerator of the fraction indicates the pitch used to obtain the tooth thickness and the number of teeth, and the denominator indicates the pitch used to determine the depth of the tooth. The clearance is greater than in the ordinary gear tooth system. The clearance angle is 20°.

Nuttall Stub Tooth—In this system, tooth dimensions are based directly on the circular pitch. The addendum is equal to 0.250 times the circular pitch, and the dedendum is equal to 0.300 times the circular pitch. The pressure angle for this system is 20°.

GEAR CUTTING OPERATIONS

Formerly, all gears were cut on milling machines, a cutter being used to produce the correct tooth shape. Then the milling machine was replaced with automatic gear cutting machines, especially for production work.

In gear cutting on the milling machine, the first problem for the machinist is to select the proper cutter because the shape of the tooth changes with the number of gear teeth. For example, the exact shape of the tooth of a 179-tooth gear is slightly different from that of a gear with 180 teeth. Although the difference is extremely small in this instance, it would be a much greater difference in gears of 20 and 21 teeth. The difference in shape is more marked in gears that have a relatively smaller number of teeth. For most practical purposes, these variations in shape can be ignored.

Cutting Spur Gears

Spur gears have straight teeth, cut parallel with the axis of rotation. Standard involute gear cutters of the arbor-mounted type are usually used to mill spur gears.

Selection of cutter—Standard sets of cutters are made by the manufacturers. For example, the Cincinnati Milacron Co. and others make eight cutters as follows:

No. 1 .... 135 teeth to a rack
No. 2 ....55–134 teeth
No. 3 ....35–54 teeth
No. 4 ....26–34 teeth
No. 5 ....21–25 teeth
No. 6 ....17–20 teeth

No. 7 .... 14–16 teeth
No. 8 .... 12 and 13 teeth

For a finer division of the number of teeth that can be cut, the Brown & Sharpe Co. manufactures cutters in half-numbers as follows:

No. 1½ .... 80–134 teeth
No. 2½ .... 42–54 teeth
No. 3½ .... 30–34 teeth
No. 4½ .... 23–25 teeth
No. 5½ .... 19 and 20 teeth
No. 6½ .... 15 and 16 teeth
No. 7½ .... 13 teeth

To select the proper cutter, it is necessary to know the number of teeth and either the diametral or the circular pitch. For example, to cut a gear with 48 teeth, first note that a No. 3 cutter can be used to cut gears with 35 to 54 teeth. The shape of the tooth that is produced will not be entirely accurate, but it will be sufficiently accurate for uses in which high speeds and smoothness in running are not essential. If the gears must be more accurate, a special cutter made to the correct shape for a given number of teeth must be used. Cutter manufacturers can furnish these special cutters.

It is not advisable to install an automatic gear cutting machine in a shop unless there is a large amount of gear cutting work to be performed. A milling machine can cut gears as rapidly as an automatic gear cutter, and it requires no more time to set up.

The chief advantage of the automatic gear cutting machine, as compared to the milling machine, lies in the fact that it can automatically index the gear and return the cutter at a rapid rate. The gear cutter operates automatically after the cut has been started; the milling machine requires the attention of the operator for indexing and advancing the work and for throwing in the feed. Even shops that have the automatic gear cutters frequently use the milling machine for odd jobs of gear cutting because it lends itself to rapid setup with no special preparation for indexing.

Setup—The gear bank is placed on a mandrel held between centers on a universal dividing head (Fig. 5-5). The dividing head is located on the table of a universal milling machine.

Fig. 5-5. Using the high-number index plate to mill a spur gear.

In gear cutting, it is especially important that there is no backlash in the indexing mechanism. The index pin should be brought around in the direction that the indexing will be done, which is preferably a clockwise direction, and permitted to drop into one of the holes.

Then set the sector for proper spacing, tighten the spindle clamp at the rear of the dividing head, and start the machine. The work should be elevated carefully until the revolving cutter barely touches the work; then set the elevating dial at "zero," run the table to the right-hand side until the cutter is cleared, and elevate for the proper depth, as indicated on the dial.

Disengage the elevating crank to reduce the possibility of disturbing the setting; then the setup is complete for beginning the milling operation. In adjusting the depth, make certain that backlash is removed before the final setting is made.

Measurement—It is customary to take a trial cut on each side of one tooth just enough to mill the full outline of the tooth. The machine is then stopped and the tooth thickness measured with a vernier gear-tooth caliper (see Fig. 5-6).

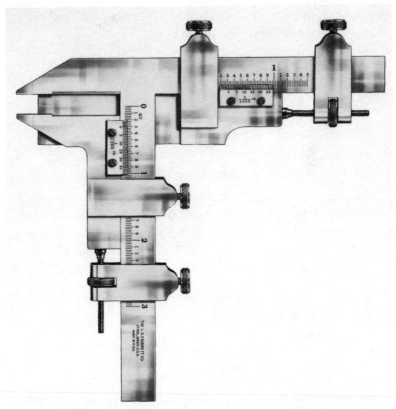

Fig. 5-6. Vernier gear-tooth caliper.

The gear-tooth vernier caliper measures chordal thickness, or thickness at the pitch line of a gear tooth, to one-thousandth of an inch. Its construction combines in one tool the function of both vernier depth gage and vernier caliper. The vertical slide is set to depth by means of its vernier plate fine-adjusting nut so that when it rests on top of the gear tooth, the caliper jaws will be correctly positioned to measure across the pitch line of the gear tooth. The

115

horizontal slide is then used to obtain the chordal thickness of the gear tooth by means of its vernier slide fine-adjusting nut.

The procedure for reading these gages is exactly the same as for vernier calipers. It is necessary to determine the correct chordal thickness and chordal addendum (or "corrected addendum"). These measurements are illustrated in Fig. 5-7.

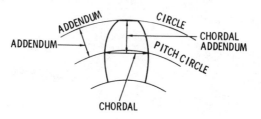

Fig. 5-7. Measuring the chordal thickness and chordal addendum.

The chordal, or straight-line, thickness of a standard gear tooth can be found by the following formula where T_c = chordal thickness; D = pitch diameter; and N = number of teeth:

$$T_c = D \sin \frac{90°}{N}$$

Example: A gear has 15 teeth and a pitch diameter of 5 inches. What is the chordal thickness?

$$T_c = 5 \sin \frac{90°}{15} = 5 \sin 6°$$

$$T_c = 5 \times 0.10453 = 0.5226 \text{ inch}$$

In measuring the chordal thickness, the vertical scale of the gear-tooth caliper is set to the chordal or "corrected" addendum to locate the caliper jaws at the pitch line. The formula used in determining the chordal addendum when a_c = chordal addendum, a = addendum, and T = circular thickness of tooth at pitch diameter D is:

$$a_c = a + \frac{T^2}{4D}$$

Example: A gear has a 1.75-inch pitch diameter, a tooth

thickness of 0.2176, and an addendum of 0.1542. What is the chordal addendum?

$$a_c = 0.1542 + \frac{0.2176^2}{4 + 1.75}$$

$$a_c = 0.1610 \text{ inch}$$

Chordal thicknesses and chordal addenda can be secured by using Table 5-1. Fig. 5-8 shows the correct way to hold a vernier gear-tooth caliper.

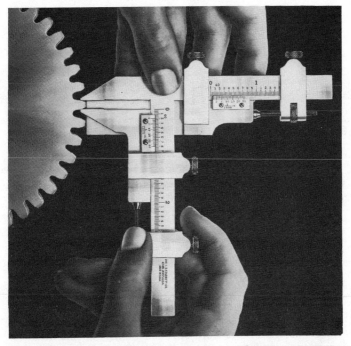

Courtesy L.S. Starrett Co.

Fig. 5-8. Correct use of the vernier gear-tooth caliper.

Cutting Bevel Gears

Bevel gears are conical gears (cone shaped) and are used to connect shafts that have intersecting axes. Hypoid gears are similar to bevel gears in their general form, but they operate on axes that are offset. Most bevel gears can be classified as either the

117

Table 5-1. Chordal Thickness of Gear Teeth
Bases of 1 Diametral Pitch

S = Module or addendum, or distance from top to pitch line of tooth
s" = Corrected S = H + S
t" = Chordal thickness of tooth
H = Height of arc

When using gear-tooth Vernier caliper to measure coarse pitch gear teeth, the chordal thickness t" must be known, since t" is less than the regular thickness AB measured on the pitch line. In referring to the table below, note that height of arc H has been added to the addendum S, the corrected figures to use being found in column s".

For any other pitch, divide figures in table by the required pitch.

No. of Teeth	t"	s"	No. of Teeth	t"	s"	No. of Teeth	t"	s"
6	1.5529	1.1022	51	1.5706	1.0121	96	1.5707	1.0064
7	1.5568	1.0873	52	1.5706	1.0119	97	1.5707	1.0064
8	1.5607	1.0769	53	1.5706	1.0117	98	1.5707	1.0063
9	1.5628	1.0684	54	1.5706	1.0114	99	1.5707	1.0062
10	1.5643	1.0616	55	1.5706	1.0112	100	1.5707	1.0061
11	1.5654	1.0559	56	1.5706	1.0110	101	1.5707	1.0061
12	1.5663	1.0514	57	1.5706	1.0108	102	1.5707	1.0060
13	1.5670	1.0474	58	1.5706	1.0106	103	1.5707	1.0060
14	1.5675	1.0440	59	1.5706	1.0105	104	1.5707	1.0059
15	1.5679	1.0411	60	1.5706	1.0102	105	1.5707	1.0059
16	1.5683	1.0385	61	1.5706	1.0101	106	1.5707	1.0058
17	1.5686	1.0362	62	1.5706	1.0100	107	1.5707	1.0058
18	1.5688	1.0342	63	1.5706	1.0098	108	1.5707	1.0057
19	1.5690	1.0324	64	1.5706	1.0097	109	1.5707	1.0057
20	1.5692	1.0308	65	1.5706	1.0095	110	1.5707	1.0056
21	1.5694	1.0294	66	1.5706	1.0094	111	1.5707	1.0056
22	1.5695	1.0281	67	1.5706	1.0092	112	1.5707	1.0055
23	1.5696	1.0268	68	1.5706	1.0091	113	1.5707	1.0055
24	1.5697	1.0257	69	1.5707	1.0090	114	1.5707	1.0054
25	1.5698	1.0247	70	1.5707	1.0088	115	1.5707	1.0054
26	1.5698	1.0237	71	1.5707	1.0087	116	1.5707	1.0053
27	1.5699	1.0228	72	1.5707	1.0086	117	1.5707	1.0053
28	1.5700	1.0220	73	1.5707	1.0085	118	1.5707	1.0053
29	1.5700	1.0213	74	1.5707	1.0084	119	1.5707	1.0052
30	1.5701	1.0208	75	1.5707	1.0083	120	1.5707	1.0052
31	1.5701	1.0199	76	1.5707	1.0081	121	1.5707	1.0051
32	1.5702	1.0193	77	1.5707	1.0080	122	1.5707	1.0051
33	1.5702	1.0187	78	1.5707	1.0079	123	1.5707	1.0050
34	1.5702	1.0181	79	1.5707	1.0078	124	1.5707	1.0050
35	1.5702	1.0176	80	1.5707	1.0077	125	1.5707	1.0049
36	1.5703	1.0171	81	1.5707	1.0076	126	1.5707	1.0049
37	1.5703	1.0167	82	1.5707	1.0075	127	1.5707	1.0049
38	1.5703	1.0162	83	1.5707	1.0074	128	1.5707	1.0048
39	1.5704	1.0158	84	1.5707	1.0074	129	1.5707	1.0048
40	1.5704	1.0154	85	1.5707	1.0073	130	1.5707	1.0047
41	1.5704	1.0150	86	1.5707	1.0072	131	1.5708	1.0047
42	1.5704	1.0147	87	1.5707	1.0071	132	1.5708	1.0047
43	1.5705	1.0143	88	1.5707	1.0070	133	1.5708	1.0047
44	1.5705	1.0140	89	1.5707	1.0069	134	1.5708	1.0046
45	1.5705	1.0137	90	1.5707	1.0068	135	1.5708	1.0046
46	1.5705	1.0134	91	1.5707	1.0068	150	1.5708	1.0045
47	1.5705	1.0131	92	1.5707	1.0067	250	1.5708	1.0025
48	1.5705	1.0129	93	1.5707	1.0067	Rack	1.5708	1.0000
49	1.5705	1.0126	94	1.5707	1.0066			
50	1.5705	1.0123	95	1.5707	1.0065			

Courtesy L. S. Starrett Co.

straight-tooth type or the *curved-tooth* type. Spiral bevels, Zerol bevels, and hypoid gears are all classified as curved-tooth gears. See Table 5-2 for number of teeth in the gear and the mating pinion.

Straight bevel gears are the most commonly used type of all the bevel gears. The teeth are straight, but their sides are tapered, so that they would intersect the axis at a common point called the pitch cone apex if they were extended inwardly. In most straight bevel gears, the face cone elements are made parallel to the root cone elements of the mating gear to obtain uniform clearance along the length of the teeth. Therefore, the face cone elements intersect the axis at a point inside the pitch cone. The calculations for the straight bevel gears are the easiest to perform; they are also the most economical of the types of bevel gears to produce.

Straight bevel gear teeth can be generated either for full contact or for localized contact. Gear teeth developed for localized contact are slightly convex in a lengthwise direction so that some adjustment of the gears during assembly is possible, and small displacements caused by load deflections can take place without an undesirable load concentration on the ends of the teeth. The slight lengthwise rounding of the sides of the teeth does not have to be computed in the tooth design, but it is taken care of automatically in the cutting operations performed on the newer bevel gear generators. Fig. 5-9 shows a set of straight bevel gears (a gear and a pinion) being power tested. These have been generated for localized contact, and the dark spots on each tooth indicates the contact area.

Zerol bevel gears have curved teeth, but they lie in the same general direction as the teeth of straight bevel gears. These bevel gears can be considered to be spiral bevel gears with zero spiral angle, and they are manufactured on the same machines as spiral bevel gears. In Zerol bevel gears, the face cone elements do not pass through the pitch cone apex, but they are approximately parallel to the root cone elements of the mating gear to provide uniform tooth clearance. Zerol bevel gears can be used when highly accurate (produced by grinding) hardened bevel gears are required.

Spiral bevel gears have curved oblique teeth. Contact begins gradually and continues smoothly from end to end; they mesh

119

Table 5-2. Bevel Gearing

Numbers of Formed Cutters Used to Mill Teeth in Mating Bevel Gear
and Pinion with Shafts at Right Angles
(Number of cutter for gear given first, followed by number for pinion.)

	Number of Teeth in Pinion																
	12	13	14	15	16	17	18	19	20	21	22	23	24	25	26	27	28
12	7-7																
13	6-7	6-6															
14	5-7	6-6	6-6														
15	5-7	5-6	5-6	5-5													
16	4-7	5-7	5-6	5-6	5-5												
17	4-7	4-7	4-6	5-6	5-5	5-5											
18	4-7	4-7	4-6	4-6	4-5	4-5	5-5										
19	3-7	4-7	4-6	4-6	4-6	4-5	4-5	4-4									
20	3-7	3-7	4-6	4-6	4-6	4-5	4-5	4-4	4-4								
21	3-8	3-7	3-7	3-6	4-6	4-5	4-5	4-5	4-4	4-4							
22	3-8	3-7	3-7	3-6	3-6	3-5	4-5	4-5	4-4	4-4	4-4						
23	3-8	3-7	3-7	3-6	3-6	3-5	3-5	3-5	3-4	4-4	4-4	4-4					
24	3-8	3-7	3-7	3-6	3-6	3-6	3-5	3-5	3-4	3-4	3-4	3-4	4-4				
25	2-8	2-7	3-7	3-6	3-6	3-6	3-5	3-5	3-5	3-4	3-4	3-4	4-4	3-3			
26	2-8	2-7	3-7	3-6	3-6	3-6	3-5	3-5	3-5	3-4	3-4	3-4	3-4	3-3	3-3		
27	2-8	2-7	2-7	2-6	3-6	3-6	3-5	3-5	3-5	3-4	3-4	3-4	3-4	3-4	3-3	3-3	
28	2-8	2-7	2-7	2-6	2-6	3-6	3-5	3-5	3-5	3-4	3-4	3-4	3-4	3-4	3-3	3-3	3-3
29	2-8	2-7	2-7	2-7	2-6	2-6	3-5	3-5	3-5	3-4	3-4	3-4	3-4	3-4	3-3	3-3	3-3
30	2-8	2-7	2-7	2-7	2-6	2-6	2-5	2-5	3-5	3-5	3-4	3-4	3-4	3-4	3-4	3-3	3-3
31	2-8	2-7	2-7	2-7	2-6	2-6	2-6	2-5	2-5	2-5	3-4	3-4	3-4	3-4	3-4	3-3	3-3
32	2-8	2-7	2-8	2-7	2-6	2-6	2-6	2-5	2-5	2-5	2-4	2-4	3-4	3-4	3-4	3-3	3-3
33	2-8	2-8	2-7	2-7	2-6	2-6	2-6	2-5	2-5	2-5	2-4	2-4	2-4	3-4	3-4	3-4	3-3
34	2-8	2-8	2-7	2-7	2-6	2-6	2-6	2-5	2-5	2-5	2-4	2-4	2-4	2-4	2-4	3-4	3-3
35	2-8	2-8	2-7	2-7	2-6	2-6	2-6	2-5	2-5	2-5	2-4	2-4	2-4	2-4	2-4	2-4	2-3
36	2-8	2-8	2-7	2-7	2-6	2-6	2-6	2-5	2-5	2-5	2-5	2-4	2-4	2-4	2-4	2-4	2-3
37	2-8	2-8	2-7	2-7	2-6	2-6	2-6	2-5	2-5	2-5	2-5	2-4	2-4	2-4	2-4	2-4	2-3
38	2-8	2-8	2-7	2-7	2-6	2-6	2-6	2-5	2-5	2-5	2-5	2-4	2-4	2-4	2-4	2-4	2-4
39	2-8	2-8	2-7	2-7	2-6	2-6	2-6	2-5	2-5	2-5	2-5	2-4	2-4	2-4	2-4	2-4	2-4
40	1-8	2-8	2-7	2-7	2-6	2-6	2-6	2-5	2-5	2-5	2-5	2-4	2-4	2-4	2-4	2-4	2-4
41	1-8	1-8	2-7	2-7	2-6	2-6	2-6	2-6	2-5	2-5	2-5	2-4	2-4	2-4	2-4	2-4	2-4
42	1-8	1-8	2-7	2-7	2-6	2-6	2-6	2-6	2-5	2-5	2-5	2-5	2-4	2-4	2-4	2-4	2-4
43	1-8	1-8	1-7	2-7	2-6	2-6	2-6	2-6	2-6	2-5	2-5	2-5	2-5	2-4	2-4	2-4	2-4
44	1-8	1-8	1-7	1-7	2-6	2-6	2-6	2-6	2-6	2-5	2-5	2-5	2-5	2-4	2-4	2-4	2-4
45	1-8	1-8	1-7	1-7	1-6	2-6	2-6	2-6	2-6	2-5	2-5	2-5	2-5	2-4	2-4	2-4	2-4
46	1-8	1-8	1-7	1-7	1-7	2-6	2-6	2-6	2-5	2-5	2-5	2-5	2-4	2-4	2-4	2-4	2-4
47	1-8	1-8	1-7	1-7	1-7	1-6	2-6	2-6	2-5	2-5	2-5	2-5	2-4	2-4	2-4	2-4	2-4
48	1-8	1-8	1-7	1-7	1-7	1-6	1-6	2-6	2-5	2-5	2-5	2-5	2-4	2-4	2-4	2-4	2-4
49	1-8	1-8	1-7	1-7	1-7	1-6	1-6	1-6	2-5	2-5	2-5	2-5	2-4	2-4	2-4	2-4	2-4
50	1-8	1-8	1-7	1-7	1-7	1-6	1-6	1-6	2-5	2-5	2-5	2-5	2-4	2-4	2-4	2-4	2-4
51	1-8	1-8	1-7	1-7	1-7	1-6	1-6	1-6	1-5	2-5	2-5	2-5	2-4	2-4	2-4	2-4	2-4
52	1-8	1-8	1-7	1-7	1-7	1-6	1-6	1-6	1-5	1-5	2-5	2-5	2-4	2-4	2-4	2-4	2-4
53	1-8	1-8	1-7	1-7	1-7	1-6	1-6	1-6	1-5	1-5	1-5	2-5	2-4	2-4	2-4	2-4	2-4
54	1-8	1-8	1-7	1-7	1-7	1-6	1-6	1-6	1-5	1-5	1-5	1-5	2-4	2-4	2-4	2-4	2-4
55	1-8	1-8	1-7	1-7	1-7	1-6	1-6	1-6	1-5	1-5	1-5	1-5	1-4	2-4	2-4	2-4	2-4

(Left axis label: Number of Teeth in Gear)

Courtesy Machinery's Handbook, The Industrial Press

Table 5-2. Bevel Gearing (Continued)

Numbers of Formed Cutters Used to Mill Teeth in Mating Bevel Gear
and Pinion with Shafts at Right Angles (continued)
(Number of cutter for gear given first, followed by number for pinion.)

		Number of Teeth in Pinion															
		12	13	14	15	16	17	18	19	20	21	22	23	24	25	26	27	28
Number of Teeth in Gear	56	1-8	1-8	1-7	1-7	1-6	1-6	1-6	1-6	1-5	1-5	1-5	1-5	1-4	1-4	2-4	2-4	2-4
	57	1-8	1-8	1-7	1-7	1-6	1-6	1-6	1-6	1-5	1-5	1-5	1-5	1-4	1-4	1-4	2-4	2-4
	58	1-8	1-8	1-7	1-7	1-6	1-6	1-6	1-6	1-5	1-5	1-5	1-5	1-4	1-4	1-4	1-4	2-4
	59	1-8	1-8	1-7	1-7	1-6	1-6	1-6	1-6	1-5	1-5	1-5	1-5	1-5	1-4	1-4	1-4	1-4
	60	1-8	1-8	1-7	1-7	1-6	1-6	1-6	1-6	1-5	1-5	1-5	1-5	1-5	1-4	1-4	1-4	1-4
	61	1-8	1-8	1-7	1-7	1-6	1-6	1-6	1-6	1-5	1-5	1-5	1-5	1-5	1-4	1-4	1-4	1-4
	62	1-8	1-8	1-7	1-7	1-6	1-6	1-6	1-6	1-5	1-5	1-5	1-5	1-5	1-4	1-4	1-4	1-4
	63	1-8	1-8	1-7	1-7	1-6	1-6	1-6	1-6	1-5	1-5	1-5	1-5	1-5	1-4	1-4	1-4	1-4
	64	1-8	1-8	1-7	1-7	1-6	1-6	1-6	1-6	1-6	1-5	1-5	1-5	1-5	1-4	1-4	1-4	1-4
	65	1-8	1-8	1-7	1-7	1-7	1-6	1-6	1-6	1-6	1-5	1-5	1-5	1-5	1-4	1-4	1-4	1-4
	66	1-8	1-8	1-7	1-7	1-7	1-6	1-6	1-6	1-6	1-5	1-5	1-5	1-5	1-4	1-4	1-4	1-4
	67	1-8	1-8	1-7	1-7	1-7	1-6	1-6	1-6	1-6	1-5	1-5	1-5	1-5	1-4	1-4	1-4	1-4
	68	1-8	1-8	1-7	1-7	1-7	1-6	1-6	1-6	1-6	1-5	1-5	1-5	1-5	1-4	1-4	1-4	1-4
	69	1-8	1-8	1-7	1-7	1-7	1-6	1-6	1-6	1-6	1-5	1-5	1-5	1-5	1-4	1-4	1-4	1-4
	70	1-8	1-8	1-7	1-7	1-7	1-6	1-6	1-6	1-6	1-5	1-5	1-5	1-5	1-4	1-4	1-4	1-4
	71	1-8	1-8	1-7	1-7	1-7	1-6	1-6	1-6	1-6	1-5	1-5	1-5	1-5	1-4	1-4	1-4	1-4
	72	1-8	1-8	1-7	1-7	1-7	1-6	1-6	1-6	1-6	1-5	1-5	1-5	1-5	1-4	1-4	1-4	1-4
	73	1-8	1-8	1-7	1-7	1-7	1-6	1-6	1-6	1-6	1-5	1-5	1-5	1-5	1-4	1-4	1-4	1-4
	74	1-8	1-8	1-7	1-7	1-7	1-6	1-6	1-6	1-6	1-5	1-5	1-5	1-5	1-4	1-4	1-4	1-4
	75	1-8	1-8	1-7	1-7	1-7	1-6	1-6	1-6	1-6	1-5	1-5	1-5	1-5	1-4	1-4	1-4	1-4
	76	1-8	1-8	1-7	1-7	1-7	1-6	1-6	1-6	1-6	1-5	1-5	1-5	1-5	1-4	1-4	1-4	1-4
	77	1-8	1-8	1-7	1-7	1-7	1-6	1-6	1-6	1-6	1-5	1-5	1-5	1-5	1-4	1-4	1-4	1-4
	78	1-8	1-8	1-7	1-7	1-7	1-6	1-6	1-6	1-6	1-5	1-5	1-5	1-5	1-4	1-4	1-4	1-4
	79	1-8	1-8	1-7	1-7	1-7	1-6	1-6	1-6	1-6	1-5	1-5	1-5	1-5	1-4	1-4	1-4	1-4
	80	1-8	1-8	1-7	1-7	1-7	1-6	1-6	1-6	1-6	1-5	1-5	1-5	1-5	1-4	1-4	1-4	1-4
	81	1-8	1-8	1-7	1-7	1-7	1-6	1-6	1-6	1-6	1-5	1-5	1-5	1-5	1-4	1-4	1-4	1-4
	82	1-8	1-8	1-7	1-7	1-7	1-6	1-6	1-6	1-6	1-5	1-5	1-5	1-5	1-4	1-4	1-4	1-4
	83	1-8	1-8	1-7	1-7	1-7	1-6	1-6	1-6	1-6	1-5	1-5	1-5	1-5	1-4	1-4	1-4	1-4
	84	1-8	1-8	1-7	1-7	1-7	1-6	1-6	1-6	1-6	1-5	1-5	1-5	1-5	1-4	1-4	1-4	1-4
	85	1-8	1-8	1-7	1-7	1-7	1-6	1-6	1-6	1-6	1-5	1-5	1-5	1-5	1-4	1-4	1-4	1-4
	86	1-8	1-8	1-7	1-7	1-7	1-6	1-6	1-6	1-6	1-5	1-5	1-5	1-5	1-4	1-4	1-4	1-4
	87	1-8	1-8	1-7	1-7	1-7	1-6	1-6	1-6	1-6	1-5	1-5	1-5	1-5	1-4	1-4	1-4	1-4
	88	1-8	1-8	1-7	1-7	1-7	1-6	1-6	1-6	1-6	1-5	1-5	1-5	1-5	1-4	1-4	1-4	1-4
	89	1-8	1-8	1-7	1-7	1-7	1-6	1-6	1-6	1-6	1-5	1-5	1-5	1-5	1-4	1-4	1-4	1-4
	90	1-8	1-8	1-7	1-7	1-7	1-6	1-6	1-6	1-6	1-5	1-5	1-5	1-5	1-4	1-4	1-4	1-4
	91	1-8	1-8	1-7	1-7	1-7	1-6	1-6	1-6	1-6	1-5	1-5	1-5	1-5	1-4	1-4	1-4	1-4
	92	1-8	1-8	1-7	1-7	1-7	1-6	1-6	1-6	1-6	1-5	1-5	1-5	1-5	1-4	1-4	1-4	1-4
	93	1-8	1-8	1-7	1-7	1-7	1-6	1-6	1-6	1-6	1-5	1-5	1-5	1-5	1-4	1-4	1-4	1-4
	94	1-8	1-8	1-7	1-7	1-7	1-6	1-6	1-6	1-6	1-5	1-5	1-5	1-5	1-4	1-4	1-4	1-4
	95	1-8	1-8	1-7	1-7	1-7	1-6	1-6	1-6	1-6	1-5	1-5	1-5	1-5	1-4	1-4	1-4	1-4
	96	1-8	1-8	1-7	1-7	1-7	1-6	1-6	1-6	1-6	1-5	1-5	1-5	1-5	1-4	1-4	1-4	1-4
	97	1-8	1-8	1-7	1-7	1-7	1-6	1-6	1-6	1-6	1-5	1-5	1-5	1-5	1-4	1-4	1-4	1-4
	98	1-8	1-8	1-7	1-7	1-7	1-6	1-6	1-6	1-6	1-5	1-5	1-5	1-5	1-4	1-4	1-4	1-4
	99	1-8	1-8	1-7	1-7	1-7	1-6	1-6	1-6	1-6	1-5	1-5	1-5	1-5	1-4	1-4	1-4	1-4
	100	1-8	1-8	1-7	1-7	1-7	1-6	1-6	1-6	1-6	1-5	1-5	1-5	1-5	1-4	1-4	1-4	1-4

Courtesy Machinery's Handbook, The Industrial Press

Courtesy Illinois Gear

Fig. 5-9. Straight bevel gears under test.

with a rolling contact that is similar to straight bevel gears. As a result of their overlapping tooth action, spiral gears transmit motion more smoothly than either straight bevel or Zerol bevel gears, thereby reducing noise and vibration, which becomes especially noticeable at high speeds. Localized tooth contact promotes smooth, quiet-running spiral bevel gears and permits some deflection in mounting without concentrating the load near either end of the tooth. One of the advantages of spiral bevel gears is the complete control of localized tooth contact. The amount of surface over which tooth contact takes place can be changed to suit the specific requirements of each job by making a slight change in the radii of the curvature of the matching tooth surfaces.

As their tooth surfaces can be ground, spiral bevel gears are advantageous in applications that require hardened gears of high accuracy. The bottoms of the tooth spaces and the tooth profiles can be ground simultaneously. This results in a smooth blending of the tooth profile, the tooth fillet, and the bottom of the tooth space. This is important from the standpoint of strength because it elimi-

122

Fig. 5-10. Lapping spiral bevel gears.

nates cutter marks and other surface irregularities that often result in stress concentrations. In some cases the gears are lapped to improve the accuracy and finish. Fig 5-10 shows the lapping of intermediate size spiral bevel gears.

Hypoid gears resemble spiral bevel gears in general appearance

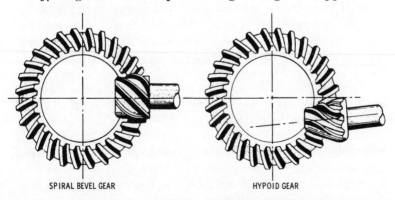

SPIRAL BEVEL GEAR   HYPOID GEAR

Fig. 5-11. Comparison of the spiral bevel gears (left) and hypoid gears (right), showing offset axis in the hypoid gears.

123

except that the axis of the pinion is offset in relation to the gear axis (Fig. 5-11). If the offset is sufficient, the shafts may pass one another, thus permitting a compact straddle mounting on the gear and pinion. As a spiral bevel pinion has equal pressure angles and symmetrical profile curvatures on both sides of the teeth, a hypoid pinion that is properly conjugate to a mating gear having equal pressure angles on both sides of the teeth must have nonsymmetrical profile curvatures for proper tooth action. To obtain equal arcs of motion for both sides of the teeth, it is necessary to use unequal pressure angles on hypoid pinions. The hypoid gears are usually designed so that the pinion has a larger spiral angle than the gear; the advantage of such design is that the pinion diameter is increased and is stronger than a corresponding spiral bevel pinion. This increment of diameter permits the use of comparatively high ratios without the pinion becoming too small to allow a bore or shank of adequate size. The sliding action along the lengthwise direction of their teeth, in hypoid gears, is a function of the difference in the spiral angles on the gear and pinion; this sliding effect makes the gears smoother running than spiral bevel gears. Hypoid gears can be ground on the same machines that are used to grind Zerol bevel gears and spiral bevel gears. Hypoid gears are frequently lapped to improve their accuracy and finish.

As mentioned previously, bevel gears are cone shaped. In form they resemble the frustums of cones. The axes of bevel gears intersect at a common apex to form an obtuse, acute, or right angle. The evolution of bevel gears is illustrated in Fig. 5-12. To draw the extended cones M and N, let the line YY represent the axis of one of the cones. Through any point on the axis YY, draw line AB at a right angle to line YY and making line AB equal to line BC. Thus, line AC is the base of the cone M.

Using any point O on line Y as the apex (depending on the slant angle desired), draw line OA and OC. Line OC represents the pitch line (corresponding to the pitch circle of a spur gear). The angle BOC is the pitch cone angle, and the distance OC is the pitch cone radius. Lay off the distance CD on the pitch line equal to the desired slant length of the frustum, and draw line DE parallel with the base line AC. The frustum of the first cone M is $ACDE$. Frustum is the part of a conical-shaped solid formed by cutting off the top by a plane parallel to the base.

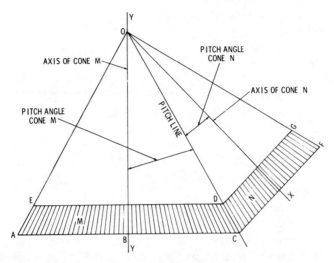

Fig. 5-12. The evolution of bevel gears from frustums of cones.

For the second cone N, draw the axis OS through point O at an angle COX equal to the pitch angle of the cone N. Also draw the line OF at an equal angle with the axis OX. From the points C and D, draw the lines CF and DG at 90° to the axis OX. Then, the frustum $CFGD$ is the frustum of the second cone N. These two frustums M and N are in contact along the common line DC, which is the common pitch line (as in the two cylinders rolling in contact); therefore, the two frustums can be considered to be two bevel gears that have an infinite number of teeth.

In actual operation the frustum N revolves about its axis OX, and the frustum M revolves about its axis OY. If power is applied to the frustum N, it will drive the frustum M by frictional contact along the common pitch line CD. The relative speed of each is inversely proportional to the diameters, or to the number of teeth in the gears. The pitch line and the pitch angle are basic lines in the laying out of bevel gears.

The following definitions relative to bevel gears are important:

1. *Pitch line*. A straight line that passes through the apex of the cone and lies in the slant surface—an element of the cone.
2. *Pitch angle*. The angle between the pitch line and the axis of the cone.

3. *Pitch cone radius.* The length of the pitch line from the apex of the cone to its base.

4. *Addendum.* The distance the tooth extends outside the pitch line at the outer edge.

5. *Dedendum.* The depth of the tooth below the pitch line at the outer edge.

6. *Addendum angle.* The angle at the apex between the pitch line and the top line of the tooth.

7. *Dedendum angle.* The angle at the apex between the pitch line and the base line of the tooth.

8. *Root angle.* The angle at the apex between the base line of the tooth and the cone axis.

9. *Force angle.* The pitch cone angle plus the addendum angle.

10. *Pitch diameter.* The diameter of the vase of the cone.

11. *Angular addendum.* The distance of the outer edge of the tooth from the cone axis less one-half the pitch diameter.

12. *Outside diameter.* The pitch diameter plus two times the angular addendum.

13. *Cutting angle.* The angle between the tooth base line and the axis of the cone.

The foregoing definitions are illustrated in Fig. 5-13. First draw the axis *YY* for the outline of the bevel gear, and generate the cone *OAC* from the given data. This determines the pitch line on either side *OA* or *OC*. Using the addendum and dedendum angles from given data, draw the top line and the base line on both sides.

Through the points *A* and *C* draw lines perpendicular to the pitch line—that is, perpendicular to lines *OA* and *OC*. The intersection of these lines with the pitch cone radius determines the outer edge of the tooth.

The inner edge of the tooth is determined from data giving the length of the tooth. The cup depth and cup clearance are laid out from the given data.

Selection of cutter—As in cutting spur gears, the first problem encountered is selection of the proper cutter for cutting the bevel gears (Fig. 5-14). The length of the teeth or the face on bevel gears is usually not more than one-third the apex distance *Ab*, and the cutters that are normally carried in stock can be used. The difference between formed cutters used for milling spur gears and those

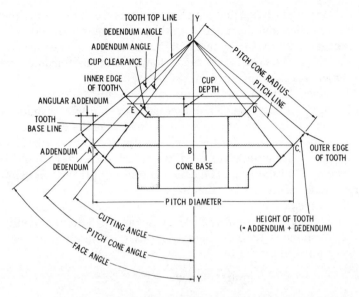

Fig. 5-13. A method of obtaining the outline of the cycloidal tooth by generating the cone (OAC) and the frustum (ACDE) of the cone, for laying out a bevel gear.

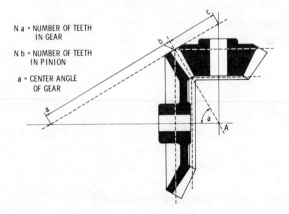

Fig. 5-14. Diagram showing the problem encountered in selection of a cutter for cutting bevel gears.

127

used for bevel gears is that bevel gear cutters are thinner since they must pass through the narrow tooth space at the small end of the bevel gear; otherwise the shape of the cutter, and hence the cutter number, are the same. If the face is longer than one-third the apex distance, special thin cutters must be used.

In selecting the cutter, measure the back cone radius *ab* for the gear, or *bc* for the pinion. This is equal to the radius of a spur gear in which the number of teeth would determine the cutter to use. Twice the back cone radius (2*ab*) multiplied by the diametral pitch gives the number of teeth for which the cutter should be selected to cut the gear. The correct number of the cutter can be determined from the list of cutters.

Example: Determine the cutter to use for cutting a bevel gear having a back cone radius *ab* of 4 inches and an 8 diametral pitch. The formula is (2*ab* × diametral pitch) = number of teeth. Substituting in the formula:

$$\text{Number of teeth} = 2\ (4) \times 8$$
$$= 8 \times 8$$
$$= 64$$

By referring to a list of standard cutters, we can select cutter No. 2 because the number of teeth (64) comes within the range (55–134) of the standard No. 2 cutter.

The number of teeth for which the cutter should be selected can also be found by the following formula:

$$\tan a = \frac{Na}{Nb}$$

To select the cutter for the number of teeth in the gear:

$$\text{Number of teeth} = \frac{Na}{\cos a}$$

To select the cutter for the number of teeth in the pinion:

$$\text{Number of teeth} = \frac{Nb}{\sin a}$$

If the gears are the same size, usually called miter gears, a single cutter can be used for both gears. Fig. 5-15 shows the power testing of large spiral miter gears. Notice that the tooth size is being

Fig. 5-15. Power-testing spiral miter gears.

measured with a vernier gear-tooth caliper. Two cutters can be required, if one gear is larger than the other.

Tables can be used to select the correct formed cutter for milling bevel gears. Table 5-2 gives the number of cutters to use for milling various numbers of teeth in the gear and pinion. This table applies only to bevel gears with axes at right angles.

Offset of the cutter—As thickness of the cutter cannot be larger than the width of the space at the small end of the teeth, it is necessary to set the cutter off center and to rotate the blank to make spaces for the correct width at the large end of the teeth. The distance to offset the cutter from the central position can be determined accurately by using the following formula:

$$\text{Offset} = \frac{T}{2} - \frac{\text{factor from table}}{P}$$

in which   P is the diametral pitch of gear to be cut,

             T is the thickness of cutter used, measured at the pitch line.

Example: It is desired to cut a bevel gear with 24 teeth, 6

Table 5-3. Table for Obtaining Set Over for Cutting Bevels

Ratio of Apex Distance to Width of Face = Apex/Face

No. of Cutter	3 / 1	3¼ / 1	3½ / 1	3¾ / 1	4 / 1	4¼ / 1	4½ / 1	4¾ / 1	5 / 1	5½ / 1	6 / 1	7 / 1	8 / 1
1	0.254	0.254	0.255	0.256	0.257	0.257	0.257	0.258	0.258	0.259	0.260	0.262	0.264
2	0.266	0.268	0.271	0.272	0.273	0.274	0.274	0.275	0.277	0.279	0.280	0.283	0.284
3	0.266	0.268	0.271	0.273	0.275	0.278	0.280	0.282	0.283	0.286	0.287	0.290	0.292
4	0.275	0.280	0.285	0.287	0.291	0.293	0.296	0.298	0.298	0.302	0.305	0.308	0.311
5	0.280	0.285	0.290	0.293	0.295	0.296	0.298	0.300	0.302	0.307	0.309	0.313	0.315
6	0.311	0.318	0.323	0.328	0.330	0.334	0.337	0.340	0.343	0.348	0.352	0.356	0.362
7	0.289	0.298	0.308	0.316	0.324	0.329	0.334	0.338	0.343	0.350	0.360	0.370	0.376
8	0.275	0.286	0.296	0.309	0.319	0.331	0.338	0.344	0.352	0.361	0.368	0.380	0.386

diametral pitch, 30° pitch cone angle, and 1¼-inch face.

Obtain the factor from Table 5-3. The ratio of the pitch cone radius to the width of the face must be determined. The pitch cone radius is equal to the pitch diameter divided by twice the sine of the pitch cone angle, which is 4 ÷ (2 × 0.5) = 4 inches. As the given face width is 1¼ inches, the ratio is 4 ÷ 1.25, or approximately 3¼ to 1. Thus, for a No. 4 cutter, the factor in the table for a 3¼ to 1 ratio is 0.280. The thickness of the cutter at the pitch line is measured by using a vernier gear tooth caliper. The cutter thickness at this depth will vary with different cutters and even with the same cutter as it is ground away in sharpening. Assuming that the thickness is 0.1745 inch, the offset can be determined by substituting in the formula:

$$\text{Offset} = \frac{T}{2} - \frac{\text{factor from table}}{P} = \frac{0.1745}{2} - \frac{0.280}{6} = 0.0406 \text{ inch}$$

An illustration of offsetting the cutter to mill bevel gears is shown in Fig. 5-16, and a method of filing to correct the shape of the tooth at the small end is shown in Fig. 5-17. A setup for milling a bevel gear on the milling machine is shown in Fig. 5-18.

In cutting bevel gears, the blank is adjusted laterally the amount of the offset, and the tooth spaces are milled around the blank. After having milled one side of each tooth, the blank is set over in the opposite direction the same amount from a position central with the cutter, and is rotated to line up the cutter with a tooth space at the small end (see Fig. 5-16). In the illustration, the table is moved in the right-hand direction, and the blank is brought to the correct position by rotating it in the direction indicated by the

Fig. 5-16. The offset of the cutter and rotation of the gear blank in cutting bevel gears so that the right-hand side of the cutter will trim the left-hand side of the gear tooth, thereby widening the large end of the tooth space.

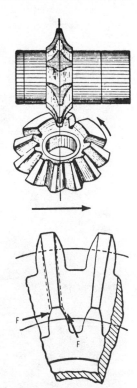

Fig. 5-17. The shape of the bevel gear tooth can be corrected at the top on the small end by filing.

arrow. A trial cut is then taken, which will leave the tooth being milled a little too thick. This trial tooth is made the proper thickness by rotating the blank toward the cutter. To test the amount of offset, measure the tooth thickness (with a vernier gear-tooth caliper) at the large and small ends. If the offset is correct, the tooth thickness will be the right size at both ends. Then the cuts can be continued until the gear is finished.

After cutting a bevel gear, the sides of the teeth at the small end should be filed as indicated by the broken lines at F in Fig. 5-17. A triangular area from the point of the tooth at the large end to the point at the small end, then down to the pitch line and back diagonally to a point at the large end will need filing.

Generating-type gear cutting equipment is extensively used in the production of bevel gears. A bevel gear tooth that is correctly formed has the same sectional shape throughout its length, but on a

131

Courtesy Cincinnati Milacron Co.

Fig. 5-18. Setup for milling a bevel gear.

uniformly diminishing scale from the large end to the small end. This correct form can be obtained only by using a generating type of bevel gear cutting machine.

If the bevel gears are too large to be cut by generating equipment (100 inches or more in diameter), they can be produced on a form-copying type of gear planer. A template is used to guide a single cutting tool in the correct path to cut the tooth profile. As the tooth profile produced by this method is dependent on the contour of the template, it is possible to produce tooth profiles to suit a variety of requirements.

Although generating methods are usually preferred, straight bevel gears are produced by milling in some instances. The milled gears cannot be produced with as much accuracy as generated gears, and generally are not suitable for use in high-speed applications or where angular motion must be transmitted accurately. Gears that are to be finished on generating equipment are sometimes first roughed out by milling. Milled gears are used chiefly as replacement gears in certain applications.

Cutting Helical Gears

Helical gears are gears that have the teeth cut along a helical surface. They are usually milled by using standard involute cutters of the arbor-mounted type on the universal knee-and-column milling machine. This permits swiveling the table to the required angle, and the workpiece can be located between centers of a dividing head and tailstock. The operation can be performed on a plain knee-and-column milling machine equipped with a universal milling attachment and a universal dividing head (Fig. 5-19). The

Courtesy Cincinnati Milacron Co.

Fig. 5-19. Milling a helical gear on a plain knee-and-column milling machine equipped with a universal milling attachment and a universal dividing head.

133

universal milling attachment permits swiveling the cutter to the required angle of swivel.

Calculations—To determine the angle to which the table must be swiveled for milling a helical gear, the following formula can be used:

$$\text{Tangent of the angle } = \frac{C}{L}.$$

in which   C is the circumference of work,
L is the lead of helix.

Example: It is desired to mill a helix with a lead of 21 inches on a gear blank 1½ inches in diameter.

Substituting the formula:

$$\text{Tangent of the angle } = \frac{4.712}{21} = 0.2244$$

On reference to a table of tangents, an angle of 12°39′ (12.64661929°) is shown to have a tangent of 0.2244. Therefore, the table angle is 12° 39′(12.64661929°).

When a helical gear is rolled on a plane surface, the traces of the teeth line up in equally spaced parallel lines (Fig. 5-20). In the illustration, the axial distance Pa between consecutive teeth is the axial pitch, Pn is the normal pitch, and Pc is the circular pitch. The lead of the helix L, the diameter of the cylinder or gear blank D, and the helix angles E and C are also shown.

Helical gears with the shafts at *right angles* to each other have different helix angles; each angle is the complement of the other. When the shafts are *parallel* (Fig. 5-21), the helix angle is the same for both gears. If the helical gears have shafts that are at an *angle less than* 90°, the sum of the helix angles is equal to the shaft angle.

Setup—Of course, the proper change gears must be selected for connecting the dividing head to the lead screw of the table, so that the workpiece will rotate as the machine table is moved horizontally (see Chapter 3 of this volume or Chapter 16 of *Machine Shop*, the second volume of the Machinists Library series).

The change gears must be placed properly to obtain the correct ratio (see Figs. 5-36 and 5-37 in Chapter 3). To generate a left-hand helix, an idler must be introduced in the gear train to reverse the

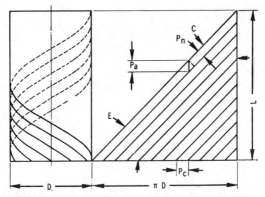

Fig. 5-20. Axial, circular, and normal pitch of equally spaced helical gear teeth.

direction of rotation. The driver and driven gears must be placed with care to cut the desired helix.

The number of the cutter to be used in milling a helical gear is also important. The calculated number of teeth to be used in selecting the proper cutter number can be found by dividing the actual number of teeth in the gear by the curve of the cosine of the helix angle.

The gear blank should be mounted on a mandrel, which in turn is mounted between centers on the dividing head. The gear blank is rotated slowly as the table advances at an angle to the axis of the cutter.

The center of the cutter should be directly above the axis of the mandrel carrying the gear blank and the arbor on which the cutter is mounted. The cutter should be centered directly above the axis of the mandrel before swiveling the table to the angle of the helix that is to be machined.

Double helical of herringbone gears are often used in parallel shaft transmissions because the opposing helices with the overlapping tooth action provide a smooth, continuous action and freedom from side thrust. These gears are useful in high-speed transmissions, such as marine reduction gears, and in connection with turbine and electric motor drives. Fig. 5-22 shows a line of continuous-tooth herringbone gear generators. Some of the gears and gear blanks are shown in the foreground.

135

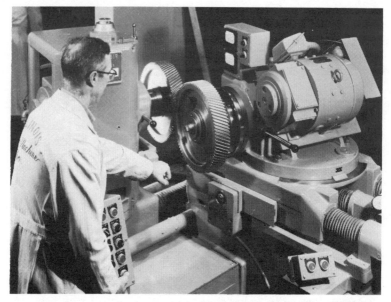

Fig. 5-21. Power inspection of helical gears with shafts parallel on a Gleason Universal Tester.

Cutting Rack Teeth

As the pitch line of the rack is a straight line, the base circle (which is parallel) becomes a straight line. The involute of the base circle is a straight line that is perpendicular to the line of action or obliquity. Therefore, the involute rack teeth have straight sides from the bottom of the tooth to the pitch line. As involute racks are to mesh with pinions that have a relatively small number of teeth, the upper part of each tooth face of the rack is rounded off through a distance equal to one-half the addendum by an arc whose radius is equal to 2.1 inches divided by the diametral pitch. This avoids interference of the teeth. Rack motion is reciprocating, with the rotation of the pinion changing direction periodically. The involute rack and pinion are shown in Fig. 5-23.

Cutter selection—A No. 1 spur gear cutter of the required diametral pitch is commonly used to mill rack teeth. The rack can be compared to a spur gear that has been straightened out and fastened to a flat surface. The center-to-center distance of the rack

Fig. 5-22. A line of continuous-tooth herringbone gear generators.

teeth must equal the circular pitch of a mating gear. The No. 1 spur gear cutter is intended for spur gears varying from 135 teeth to a rack. The depth of tooth is calculated in the same manner as for spur gear teeth.

Setup—Rack teeth, either straight or inclined with respect to the rack blank, can be milled on the milling machine by means of the rack cutting and rack indexing attachments. The rack with inclined teeth is known as a spiral rack, and can be cut only on universal milling machines (Fig. 5-24) since the table can be swiveled to the required angle.

The rack blank is held in a rack vise that is clamped to the table

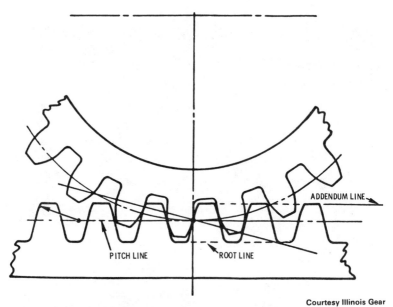

Fig. 5-23. Diagram of a rack-and-pinion gear in mesh.

of a universal milling machine (Fig. 5-25). Then the worktable is swiveled in a clockwise direction to a required angle (23°37' or 23.616666667°). This places the rack blank so that the rack teeth are parallel to the saddle cross feed and to the milling cutter mounted on the *rack cutting attachment* (see Fig. 5-25).

The machine table moves parallel to the linear pitch of the rack teeth. Thus, to space the teeth at the normal pitch, it is necessary to index each tooth into position by moving the table a distance equal to the linear pitch. The *rack indexing attachment* (see Fig. 5-25) is used to perform the indexing operation. The cutter is mounted on a rack milling attachment that places the cutter parallel to the cross feed.

In milling the spiral rack (see Fig. 5-24), the workpiece is set vertically by hand feeding the knee upward until the workpiece contacts the cutter. Then the dial on the knee adjustment is set at "zero" reading. After clearing the cutter, the knee is raised a distance equal to the whole depth of the tooth to obtain the required depth of cut.

Courtesy Cincinnati Milacron Co.

Fig. 5-24. Setup for milling the teeth of a spiral rack.

The workpiece can be located longitudinally by moving the table and saddle until it lightly contacts one side of the cutter. Then the dial of the table lead screw is set to "zero" reading. After clearing the workpiece from the cutter, the table is moved in a right-hand direction a distance approximately equal to one-half the cutter thickness. This makes the setup ready for milling the rack teeth by feeding the saddle inwardly and returning it to the starting position after each pass.

The next tooth is indexed into position by turning the crank of the rack indexing attachment until the index plate has made one complete turn. The dimensions and spacing of the first few teeth should be checked. The procedure can then be repeated several times until the required number of rack teeth has been milled.

Cutting Worm and Worm Wheel Teeth

Thread milling cutters and *hobs* can be used on the milling machine to cut worm wheel teeth. This type of gearing is used in

139

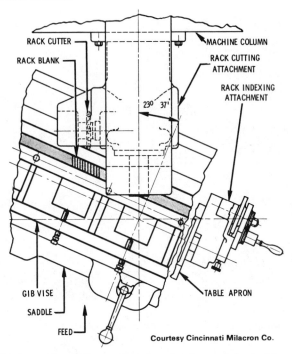

RACK CUTTER

RACK BLANK

MACHINE COLUMN

RACK CUTTING
ATTACHMENT

RACK INDEXING
ATTACHMENT

23° 37'

GIB VISE

SADDLE

FEED

TABLE APRON

Courtesy Cincinnati Milacron Co.

Fig. 5-25. Setup for milling rack teeth, using the cross feed of the machine. The milling cutter is mounted on a rack cutting attatchment that places the cutter parallel to the motion of the saddle.

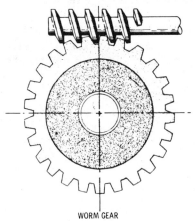

WORM GEAR

Courtesy Cincinnati Milacron Co.

Fig. 5-26. A typical worm and worm wheel.

Courtesy Cincinnati Milacron Co.

Fig. 5-27. Setup for gashing the teeth in a worm wheel.

drive arrangements to obtain a reduction in speed ratio between the worm and worm wheel (Fig. 5-26). The ratio of the drive is independent of the pitch diameters of the worm and worm wheel. The worm is the driver, and the worm wheel is the driven gear.

The *worm* is a screw with either a single thread or multiple threads; the form of the axial cross section is the same as that of a rack. The teeth of the *worm wheel* have a special form that is required to provide proper conditions for meshing with the worm.

The ratio of the drive is equal to the number of teeth in the worm wheel if the worm has a single thread. The drive ratio decreases as the number of threads in the worm is increased, with a constant number of teeth in the worm wheel. For worms with double and quadruple threads, for example, the drive ratio is one-half and one-fourth, respectively, of the number of teeth in the worm wheel.

The worm can be milled on a milling machine by means of thread milling cutters. The worm is held between centers of a dividing head and tailstock; the cutter is mounted on the spindle of a universal milling attachment and swiveled to the helix angle of the worm threads. This setup can be made on a universal knee-and-column milling machine.

Other methods of machining worm threads include: milling with a disk-shaped cutter; hobbing on a regular gear-hobbing

141

Courtesy Cincinnati Milacron Co.

Fig. 5-28. Setup for hobbing the teeth of a worm gear.

machine; generating on a worm thread generator; cutting in a lathe if the lead angle is not too great; and grinding to the finished size; after milling and hardening the worm. The latter is the preferred method if a hardened worm is desired.

Two operations are required to cut the teeth on a worm wheel. They are: (1) *gashing* the teeth and (2) *hobbing* the teeth to the finished size and shape.

In the gashing operation, the gear teeth are *roughed* with an involute gear cutter that has the same pitch and diameter as the worm. After aligning the gear blank on the center of the gashing cutter both crosswise and longitudinally, the table of the machine is swiveled to the gashing angle in a counterclockwise direction for a right-hand worm. The gashing operation is then performed by feeding the work vertically to the depth of the teeth, leaving enough stock for the finishing operation (Fig. 5-27).

If the gashing angle is not given, it can be calculated from the known value of the lead and pitch diameter of the worm. Tables are usually provided to give the gashing angles for worm wheels for a variety of worms having standard diameters and leads.

The worm wheel is held between centers for the hobbing operation, but it is free from the driving dog of the dividing head; thus the hob drives the wheel while the teeth are being cut. The axis of the worm wheel is at right angles to that of the worm; therefore, it is necessary to set the table of the universal milling machine in the usual straight position so that the axis of the worm wheel is at right angles to the arbor on which the hob is mounted (Fig. 5-28).

The table of the machine is locked in position to prevent its moving while the teeth are being hobbed. The workpiece must be adjusted so that the hob centers over the rim of the worm wheel. Then the work is raised gradually until the correct depth is obtained. If it is necessary to remove a large amount of stock or an exceptional finish is required, the worm wheel can be passed under the hob several times, bringing it to the final depth for the last revolution.

SUMMARY

A disk, or wheel, that has teeth around its periphery for the purpose of providing a positive drive by meshing the teeth with similar teeth is called a gear. The spur gear is the simplest type and is the one most commonly used. The spur gear has straight teeth that are cut parallel with the axis of rotation of the gear body. All other forms of gears, such as bevel, helical, worm, and worm wheels are modifications of the spur gear.

Various terms are used in connection with gears and gear cutting. Some of these terms are: pitch circle, pitch point, face, addendum circle, dedendum circle, circular pitch, chordal pitch, pitch diameter, and pressure angle.

The pitch circle is the reference circle of measurement. The pitch point is the point of tangency of two pitch circles, or the pitch circle and pitch line, and is located on the line centers. The face of a gear tooth is the surface of the tooth between the pitch circle and the top of the tooth. The addendum circle is the circle that passes through the top of the gear teeth and is the same as the outside diameter of the gear.

The addendum is the height of the tooth above the pitch circle, and the dedendum is the depth of the tooth space below the pitch circle. Circular pitch is the distance from the center of one tooth to the center of the adjacent tooth when measured on an arc of the pitch circle.

The setup for cutting each type of gear is quite different. The spur gear requires a rather simple setup and can be done on a milling machine. Bevel gears require more calculations and a more complicated setup that does not result in an accurate gear when cut on a mill. The helical gear requires change gears connecting the lead screw of the table to the dividing head. Worm wheels involve the use of a hob, and rack milling usually requires special rack attachments to the milling machine.

REVIEW QUESTIONS

1. What is a spur gear?
2. Name a few of the gear teeth terms and explain their meaning.
3. What is the chief difference between a stub tooth and the full-depth tooth gear?
4. What is a Zerol bevel gear?
5. What is a hypoid gear assembly?
6. What is a helical gear?
7. How is the ratio calculated for a worm and worm wheel?
8. How is a worm thread machined?
9. What is a hob?
10. Explain the use of a vernier gear-tooth caliper.

CHAPTER 6

Cams and Cam Design

A cam is a mechanical device used to convert *rotary* motion into *varying speed reciprocating* motion. This basic action results when an irregularly shaped disk revolves on a shaft and imparts irregular motion to a follower. Cams, in general, can be divided into two classes: (1) uniform motion cams and (2) uniformly accelerated motion cams.

CAM PRINCIPLES

The motion of the cam is transmitted to a follower, which remains in contact with the acting surface as the cam revolves. The follower corresponds to the plunger or lifter of the valve mechanism (Fig. 6-1).

Cams are made in two basic shapes: (1) plate cams and (2) cylindrical cams. In general, the follower of a plate cam operates

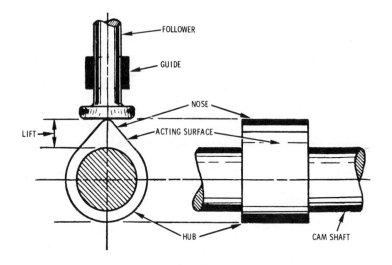

Fig. 6-1. Basic diagram of a typical cam.

perpendicular to the axis of the cam shaft while the follower of the cylindrical cam oscillates parallel with the axis of the cam shaft (Fig. 6-2).

Cam followers may have a variety of shapes. Four common shapes are: (1) knife edge or pointed, (2) roller, (3) flat-faced, and (4) spherical-faced (Fig. 6-3). Follower motion is assured by the use of gravity, by spring action, or by operating in a groove. A groove cam has the advantage of being able to apply force in opposite directions (Fig. 6-4).

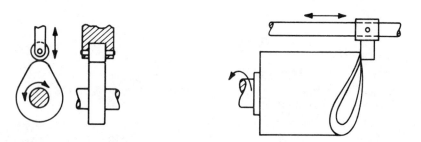

Fig. 6-2. Follower motion in respect to cam action.

146

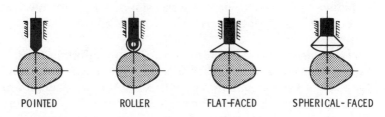

POINTED   ROLLER   FLAT-FACED   SPHERICAL-FACED

Fig. 6-3. Types of cam follower shapes.

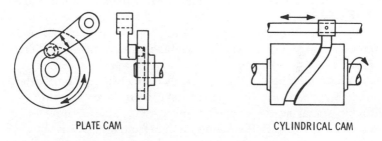

PLATE CAM   CYLINDRICAL CAM

Fig. 6-4. Groove cam applies force in opposite directions.

Uniform Motion Cams

If a body moves through equal spaces in equal intervals of time, it is said to have uniform motion; that is, its velocity is constant. The uniform motion cam moves the follower at the same rate of speed throughout the stroke, from the beginning to the end. This means that if the movement is rapid, there is a distinct shock at the beginning and end of the stroke because the movement is started from "zero" to the top speed of the uniform motion and is stopped just as abruptly. Therefore, it is important to construct cams in such a way that these sudden shocks are avoided either in starting the motion or in reversing the direction of motion of the follower. These cams are suitable for machinery operating at a slow rate of speed.

Uniformly Accelerated Motion Cams

The cam best suited for high speeds is one in which the speed is slow at the beginning of the stroke, accelerated at a uniform rate until maximum speed is attained, and then uniformly retarded

147

until the follower is stopped or nearly stopped when the reversal takes place. This type of cam is called a uniformly accelerated (or retarded) motion cam.

Harmonic motion—This form of varying or uniformly accelerated motion is illustrated in Fig. 6-5. If a point *A*, as indicated by the arrow, travels around the circumference of a circle at uniform velocity and another point *B* travels across the diameter of the circle at a variable velocity so that it is always at a point where a perpendicular from point *A* would intersect the diameter, then the point *B* increases from "zero" at the starting point *C* until it reaches the center point *O*; from that point the velocity decreases to "zero" as it reaches the point *D* at the end of its travel. The harmonic motion cam is satisfactory for machinery operating at moderate speeds.

Gravity motion—In this uniformly accelerated (and retarded) motion, the rate of acceleration or retardation bears the same ratio to the speed as the acceleration or retardation produced by gravity—hence its name. A body falling from rest travels about 16 feet during the first second. During the next second, its velocity increases by 32 feet, making 48 feet the distance covered during that second. During each succeeding second, the velocity increases by 32 feet. The increase in velocity is in ratio of 1, 3, 5, 7, 9,

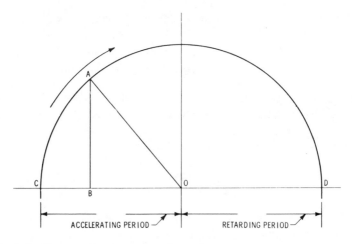

Fig. 6-5. Diagram showing harmonic motion.

148

etc. This ratio can be used in laying out a gravity motion cam. The gravity motion cam is best for machinery operating at high speeds.

How a Cam Operates

Tangential cam action in opening and closing a valve is illustrated in Fig. 6-6. In the diagram, the cam and valve positions are shown for each 22.5° during 180°, or one-half of a revolution. The successive cam positions are indicated by the numerals 1, 2, 3, etc.; the corresponding valve positions are also indicated by the same numbers. In cam positions 1 and 2, the follower is in contact with the cam hub; the valve remains seated in these positions. As can be seen in the diagram, the valve remains seated until the straight portion of the cam's acting surface attains a horizontal position, as indicated by the dotted outline A.

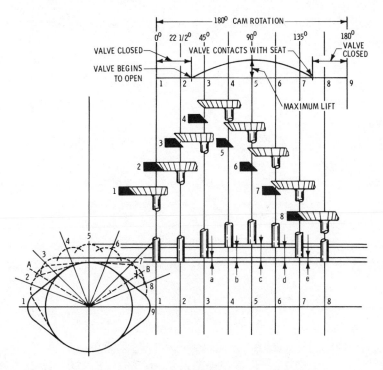

Fig. 6-6. Cam action needed when opening and closing a valve.

149

When the cam reaches position 3, it has pushed the follower and valve upward a distance a; the valve position is shown directly above. Similarly, for cam positions 4, 5, 6, etc., the valve is upward at the distances b, c, d, etc., respectively, from its seat. The valve reaches its closing point between positions 7 and 8.

The curve at the top of the diagram indicates the movement of the valve for 180°, or one-half of a revolution. The curve indicates that the cam is a type that has a varying motion; that is, the motion is accelerated in its opening and retarded in its closing movements.

The points of opening and closing the valves are the points where the acting surface is tangent to the hub surface (Fig. 6-7). In the tangential cam, the acting surface between the nose and hub is straight. Perpendiculars to the straight surface indicate the valve positions as determined by the positions of the ends of the valve stems; in the rotation of the cam, these two points cause the valve to open and close as they come in contact with the follower. The arc ARB is the rest period in which the valve remains closed.

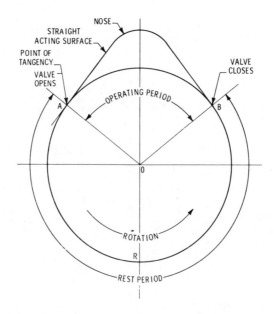

Fig. 6-7. The determination of points of opening and closing in cam rotation.

When the follower motion is parallel to the camshaft, it is necessary to use some form of a cylindrical cam. This usually consists of a roller follower operating in a groove cut in the rotating cylinder cam. The groove is cut according to the amount of displacement and the type of motion desired (Fig. 6-8).

CAM DESIGN

Cam action has its limits; under certain conditions, the result can be a noisy and fast-wearing mechanism. Some basic or fundamental facts should be considered in cam design.

Displacement diagrams—Since the motion of the follower is of primary importance, its rate of speed and its various positions should be carefully planned in a displacement diagram. A dis-

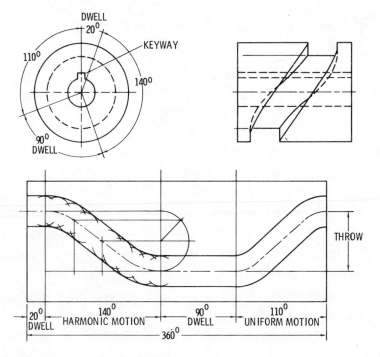

Fig. 6-8. Layout of cylindrical cam.

placement diagram is a curve showing the displacement of the follower as ordinates erected on a base line that represents one revolution of the cam (Fig. 6-9). These three diagrams illustrate the various cam motions, while Fig. 6-10 shows a displacement diagram of a cam combining all three motions. Notice that the uniform motion has been modified to reduce shock.

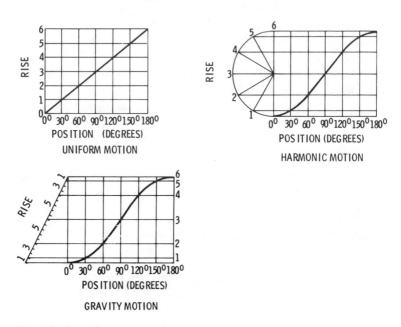

Fig. 6-9. Cam displacement diagrams.

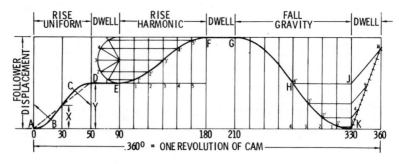

Fig. 6-10. Displacement diagram with three types of motion.

Regardless of the desirability of quick action in valve opening and closing, the opening and closing faces should meet the hub circle tangentially; that is, there should be no abrupt change in direction of the acting surface. This is especially important on the opening face because the tension of the spring opposes a sudden starting action of the valve when it begins to open. Also with the steeper opening face, the lateral thrust on the follower is greater, which increases friction and wear.

In a uniform motion cam, the contour of the action surface moves the follower at the same rate of speed from beginning to end of the valve movement. This is illustrated in the layout of a uniform motion cam (Fig. 6-11).

Valve lift is indicated in the upper part of the illustration (see Fig. 6-11). If a line is divided into a number of equal parts and if perpendiculars are drawn from the points *1, 2*, etc., to intersect the inclined line *le*, movement of the valve is indicated by the distances *2a, 3b*, etc., for equal arcs of cam rotation (*1-2, 2-3*, etc.).

In the lower part of the illustration (see Fig. 6-11), the horizontal and vertical axes are laid out. and the lines *OA* and *OB* are con-

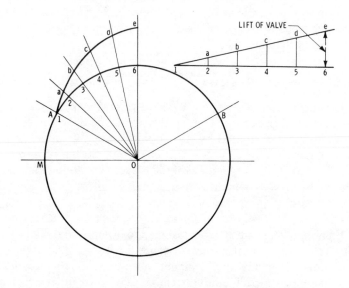

Fig. 6-11. The layout of a cam having a uniform motion.

structed so that the obtuse angle *AOB* represents the operating
angle of the cam. Then, a circle having a radius equal to that of the
cam hub is described, and the arc *1-6* is divided into the same
number of equal parts as the line *1-6* was divided in the upper part
of the illustration. The radii of the circle are extended indefinitely
through the points *1*, *2*, etc., far enough to lay off the distances *2a*,
3b, etc., equal to the corresponding distances in the upper part of
the illustration.

Then, the curve *abcde* can be drawn through the points on the
radii to represent one-half of the acting surface of the cam. The
other one-half of the curve is identical and can be obtained by
using the same construction method.

The chief disadvantage of the uniform motion cam is that valve
motion begins at full speed and stops in the same abrupt manner,
which is a distinct shock to the valve gear in starting and stopping
actions. This is illustrated in Fig. 6-11. As point *1* of the cam
contacts the follower, the follower immediately moves at full
speed, that is, without a gradual acceleration period. It is common
practice to modify this type of action to reduce shock. This can be
seen in Fig. 6-10, where the dash line represents the true uniform
motion and the solid line represents the modified uniform motion.
This type of cam shape is unsatisfactory for automobile engines
because of the high speed at which they are run.

Design for Gas Engines

Layout of a cam for operating a valve under specified conditions is not complicated. The rapidity with which the valves open
and close is determined by the slope of the valve opening and
closing faces on the cam. Steep opening and closing faces on the
cam provide quick movement of the valves. The following example can be used to illustrate the design of a uniformly accelerated
motion cam.

Example: The cam is to be designed to operate an intake valve.
The following conditions are specified: lift, ¹¹⁄₃₂ (0.34375 inch;
cam hub circle, 1½ (1.50) inches; nose arc, ⁵⁄₁₆ (0.3125) inch,
admission period, 240°; and the cam is to provide quick acceleration in the opening and closing actions of the valve.

As an admission period occupies 240° of crankshaft rotation, it will occupy only 120° of camshaft rotation because the camshaft speed is one-half that of the camshaft (240° ÷ 2 = 120°). The design of the cam with curved opening and closing faces for operation of the intake valve is illustrated in Fig. 6-12.

First draw the horizontal and vertical axes (see Fig. 6-12). Draw lines *Aa* and *Bb* through the center point *O*, so that the angle *AOB* is equal to 120° and is bisected by the vertical axis (∠*AOM* = 60° and ∠*BOM* = 60°). Describe the cam hub circle [1½ (1.50) inches], and lay off the lift [¹¹⁄₃₂ (90.34375) inch] from the point *m*. With the center of the vertical axis *OM*, describe the nose arc [with radius of ⁵⁄₁₆ (0.3125) inch] through point *n*. On line *Aa*, by trial, find the center of a circle that is tangent to both the cam hub circle and the nose arc; describe the arc connecting them to give the contour of the opening face. The contour of the closing face is identical and can be obtained by the same method of construction. This is frequently referred to as a circular arc cam since the surface

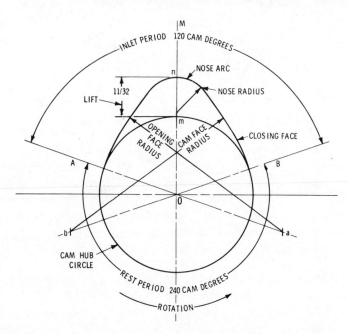

Fig. 6-12. The design of a cam having uniformly accelerated motion.

155

consists entirely of tangent arcs of circles. It has the advantages of being comparatively easy to manufacture.

Another example can be used to illustrate the design of a tangential cam for intermittent motion. Using the same lift, admission period, and cam hub diameters as in the preceding example, the problem differs in this example (Fig. 6-13) as follows: (1) the nose arc is described from the center point O of the cam; and (2) the opening and closing faces are tangent lines that are tangent to the hub circle at the points of intersection of the lines Aa and Bb with the hub circle.

It should be noted that the action on the follower is very abrupt at the points f and f' (see Fig. 6-13). This can be avoided by connecting the tangents and the nose arc with relief arcs, as shown. The amount of relief, that is, the acceleration and retardation periods, depend on the radius of the relief arcs. Solid lines are used

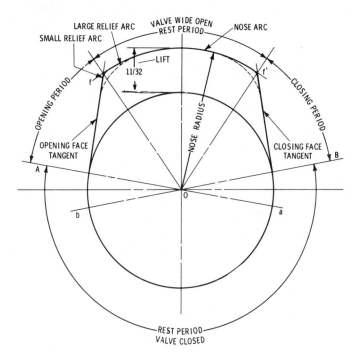

Fig. 6-13. The design of a tangential cam for intermittent motion.

to show the arcs with smaller relief, and dotted lines are used to show the arcs with larger relief. This cam is referred to as tangential cam since the opening and closing faces are tangent to the hub circle and to the relief arcs.

An overtravel cam is sometimes used to give the valve more opening than is necessary for proper distribution of the charge. Overtravel is used to increase acceleration and retardation and to obtain relief at reversal of movement. Valve lift can vary with the manufacturer.

The design of cams for exhaust valves is different from that of intake valves. The cams for exhaust valves have a more flattened nose so that a period of maximum opening is provided for the free escape of the burned gases.

A comparison of sharp-nosed cams and broad-nosed cams is illustrated in Fig. 6-14. When the cams are in the positions M and S, they are at the points where the valves begin to open; and at the positions L and F, they are fully open. As noted in the illustration, the broad-nosed cam provides greater acceleration in opening the valve.

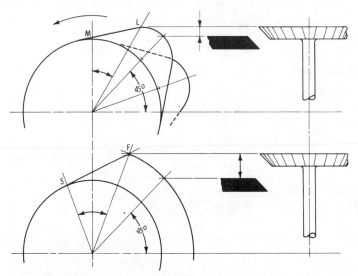

Fig. 6-14. A comparison of acceleration on a sharp-nosed cam (top) and a broadnosed cam (bottom). The broad-nosed cam provides greater acceleration in the opening of the valve.

Design for Automatic Screw Machines

The Brown & Sharpe No. 2 high-speed machine is used here to illustrate typical cam design. (The reader may wish to refer to Chapter 15 of Volume 1, which describes the setup of this machine.) A typical workpiece, shown in Fig. 6-15, is also used. This particular workpiece has a turned diameter, a formed head, and a rounded end. The necessary steps in cam design are as follows:

1. Decide on the method to be used in doing the job, tools required, and order of operations.
2. Determine spindle speed.
3. Calculate throws of the cam lobes and the spindle revolutions required for the cutting operations.
4. Overlap the operations wherever possible.
5. Figure spindle revolutions required for idle movements.
6. Provide clearance space if necessary.
7. Find total estimate spindle revolutions required to finish a piece; then select the actual revolutions available with regular change gears that come nearest the estimated number.
8. Readjust the estimated spindle revolutions to total the actual number available on the machine.
9. Calculate the hundredths of cam surface required for each operation and idle movement.

As an illustration, we can assume that the turned diameter on the workpiece (see Fig. 6-15) requires a good finish, the following operations being necessary:

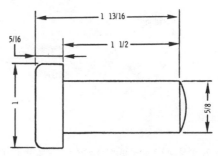

Fig. 6-15. Workpiece used to illustrate typical cam design for an automatic screw machine.

1. Rough turn to within 0.010 inch of the finished diameter.
2. Turn with a balance turning tool to within 0.010 inch of the finished length.
3. Finish turn with a box tool to the finished diameter.
4. Advance the cross slides to form the head and to sever the workpiece from the bar.
5. With the forming tool, finish under the head, removing the 0.010 inch of length left by the turning tools.
6. Round the end with the cutoff tool for the next workpiece.

A work sheet (Fig. 6-16) can be used to assemble data for making the cam. This is not justified unless there is a large volume of work to be performed. The cam design work sheet gives all the data needed to do the job with all the operations listed in the order in which they are performed.

Lever templates (Fig. 6-17) can be used to calculate close timing in cam design. It is necessary to know the relative positions of one or both of the cross slides when the turret is advancing, retreating, or at its extreme inward position. Likewise, it is essential at times to know the position of the turret slide relative to one or both cross slides. The templates are useful in determining the exact relative positions of the slides on the drawing, without having to resort to figures or calculations. They are useful in determining the amount of space to allow for tool clearance, or in determining points that correspond on the lead cam and cross slide cams so that the movements of the turret slide and cross slides can be synchronized when their combined action is required to feed a tool, as in swing tool work.

The templates pivot at the center of the cam drawing or trial sketch on a pin inserted through the hole in a button at the end of the long arm. They are made of clear material so that the lines of the cam drawing can be seen. The centers from which the lead and cross slide levers swing represent the fulcrum centers of the actual levers on the machine. The two levers are the same length as the corresponding levers on the machine, and the rounded portion at their ends is the same diameter as the rolls that run on the cams. If, for example, the lead cam lever is at the 85th division line on lobe *a* and it is desired to have the front cross slide begin to guide or feed the tool at that instant, the corresponding hundredth division line

CAM DESIGN WORK SHEET

PART NAME OR NO.....Example 1 .... 　MATERIAL BRASS ....

　SURFACE FEET FOR STOCK ....

　" 　" 　" 　THREAD ....

　" 　" 　" 　DRILL ....

　" 　" 　" 　TURN ....

SPINDLE { FORWARD....3000 ....

R/MIN 　{ BACKWARD ....

　SECONDS....12 ....

　GROSS PRODUCTION PER HOUR..300 ....

　MADE ON..No. 2 High Speed ....

Order of Operations	Throw	Feed	Spindle Revolutions: For Each Operation	Spindle Revolutions: After Deducting for Operations Overlapped	Spindle Revolutions: Readjusted to Equal Revolutions Obtainable with Regular Change Gears	Spindle Revolutions: For Spindle Revolutions in Preceding Column	Hundredths: For Spindle Revolutions in Preceding Column	Hundredths: For Operations That Are Overlapped
Feed Stock to Stop (4" long)					30	30	5	
Double Index Turret					30	30	5	
Rough Turn - No. 22								
Balance Turning Tool	1.500	0.010	150		150	150	25	
Double Index Turret					30	30	5	
Finish Turn - No. 22B								
Box Tool	1.500	0.010	150		150	150	25	
Clear						24	4	3
Form - Front Slide {	182	.003	60					10
{	010	0.001	18					3
Double Index Turret								
Cut Off - Back Slide	574	0.0035	164	164		186	31	
96 Hundredths equal				554				
Estimated Total Revs., if spindle runs 3000 r/min continually				576				
Nearest Actual Spindle Revs. available on Machine						600	100	

NOTE—A dimensioned pencil sketch of the piece is often drawn in blank space at top of this sheet.

Fig. 6-16. Cam design work sheet.

on the cross slide cam can be found by placing the lead cam lever, with the roll barely touching the line of lobe a, at the 85th division line; then swing the cross slide lever downward until it touches the arc c, which represents the bottom of the throw for lobe b. A radial line from the center point of the cam drawing to the cam circle and passing through the center of roll on the cross slide lever shows graphically the hundredth division line on which to start the lobe b—at 86.5 hundredths in this instance.

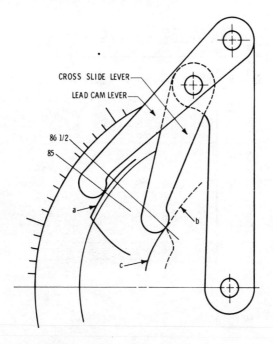

Fig. 6-17. Using lever templates in cam design.

Estimating spindle revolutions—In laying out cams, the approximate feeds and speeds can be obtained from Table 6-1. From the table it can be noted that brass can be worked at the fastest speed of the machine, which is 3000 r/min, as indicated in the table for laying out cams (Table 6-2). It should be noted in Table 6-2 that when threading to a shoulder, the required relative position of the clutch teeth at the time of reversing is obtained by using the table values. This exists when a whole number is obtained by multiply-

161

Table 6-1. Approximate Cutting Speeds and Feeds for Standard Tools

TOOL	CUT — Width or Depth	CUT — Dia. of Hole	BRASS FREE CUTTING — Feed	BRASS FREE CUTTING — Speed in Surface Feet	MATERIAL — Mild or Soft Steel .10–.20% Carbon — Feed	Mild — Speed, Carbon Tools	Mild — Speed, H.S.S. Tools	Tool Steel .80–1.00% Carbon — Feed	Tool Steel — Speed, Carbon Tools	Tool Steel — Speed, H.S.S. Tools
Boring Tools	.005		0.012	Use Maximum Spindle Speed	0.008	50	110	0.004	30	60
Box Tools—Roller Rest	1/32		0.010		0.010	70	150	0.005	40	75
1 Chip Finishing	1/16		0.008		0.008	70	150	0.004	40	75
	1/8		0.008		0.007	70	150	0.003	40	75
	3/16		0.006		0.006	70	150	0.002	40	75
	1/4		0.010		0.005	70	150	0.0015	40	75
Finishing	.005		0.003		0.010	50	110	0.006	30	75
Center Drills		under 1/8	0.006		0.0015	50	110	0.001	30	75
		over 1/8	0.0015		0.0035			0.002		
Cut-Off Tools — Angular	3/64–1/8		0.0035		0.0006	80	150	0.0004	50	85
Circular	1/16–1/8		0.0035		0.0015	80	150	0.001	50	85
Straight			0.002		0.0015	80	150	0.001	50	85
Dia. Stock under 1/8"					0.0008			0.0005		
Button Dies					0.001					
Chaser Dies					0.0014					
Drills Twist Cut		0.02	0.0014		0.002			0.0006	14	20
		0.04	0.002		0.0025	30	40	0.0008	16	45
		1/16	0.004		0.0035	40	60	0.0012	30	45
		3/32	0.006		0.004	40	60	0.0016	30	45
		1/8	0.009		0.005	40	60	0.002	30	45
		3/16	0.012		0.005	40	75	0.003	30	60
		1/4	0.014		0.006	40	75	0.0035	30	60
		5/16	0.016		0.006	40	75	0.004	30	60
		3/8	0.016		0.006	40	85	0.004	30	60
		1/2	0.016		0.0009	40	85	0.004	30	60
		5/8	0.016		0.0008	40	85	0.0006	30	60
Form Tools—Circular	1/16–1/4		0.002					0.0005		

Tool	Size		Available on Machine						
Hollow Mills (Turned diam. under 5/32")	3/8–1/2	0.0015	Available on Machine	0.0007	80	150	0.0004	50	85
Balance Turning Tools (Turned diam. over 5/32")	5/8–3/4	0.0012		0.0006	80	150	0.0003	50	85
	1	0.001		0.0005	70	150	0.008	40	85
	1/32	0.012		0.0004	70	150	0.006	40	85
	1/16	0.010		0.010	70	150	0.010	40	85
	1/32	0.017		0.009	70	150	0.008	40	85
	1/16	0.015		0.014	70	150	0.008	40	85
	1/8	0.012		0.012	70	150	0.006	40	85
	3/16	0.010		0.010	70	150	0.0045	40	85
		0.009		0.008		150	0.010	40	85
Knee Tools	1/64–1/32	0.020		0.007	150		0.008	105	
Knurl Tools — Turret	on	0.040		0.010	150		0.010	105	
	off	0.004		0.012	150		0.025	105	
Side or Swing		0.006		0.015	150		0.002	105	
Top		0.005		0.030	150		0.003	105	
		0.008		0.002	150		0.002	105	
				0.004		150	0.004	40	80
Pointing & Facing Tools		0.001		0.003	70		0.0005	40	80
Reamers & Bits	.003 to .004 — 1/8 or less	0.0025		0.006	70		0.0008	40	60
	.004 to .008 — 1/8 or over	0.010		0.0008	70	105	0.006	40	60
		0.007		0.002	70	105	0.004	40	60
		0.010		0.008	70		0.006	40	60
				0.006			0.008		60
				0.010					75
Recessing Tools End Cut		0.001		0.0006	70	150	0.0004	40	75
Inside Cut	1/16–1/8	0.005		0.003	70	150	0.002	40	60
	1/8–1/4	0.0025		0.002	70	105	0.0015	40	60
		0.0008		0.0006	70	105	0.0004	40	85
Swing Tools, Forming	3/8–1/2	0.002		0.0007	70	150	0.0005	40	85
		0.0012		0.0004	70	150	0.0003	40	85
		0.001		0.0003		150	0.0002		
Turning Straight	1/32	0.0008		0.006	70	150	0.0035	40	85
	1/16	0.008		0.004	70	150	0.003	40	85
	1/8	0.006		0.003	70	150	0.002	40	85
	3/16	0.005		0.0025	70	150	0.0015	40	85
Taps		0.004			25	30		12	15
Swing Tools									

Taper turning same as straight turning but the feed is taken slow enough for the greatest depth of cut.

Table 6-2. No. 2 Automatic High-Speed Screw Machine (With 36 Spindle Speeds)

(Begins with Machine Serial No. 7352)
Table for Laying Out Cams—Driving Shaft 240 r/min

Feed Gears — DRIVER → A; B (1st, 2nd); C

Feed gear settings (by Time in Seconds to Make One Piece):

Hundredths of Cam Surface to Feed Stock	17	15	14	13	12	12	11	11	10	10	9	9	8	8	7	6	5
Gear on C	40	42	40	48	40	54	42	60	77	60	78	70	80	72	80	77	80
2nd Gear on B	80	72	72	72	80	90	72	84	60	72	60	60	48	48	42	40	80
1st Gear on B	32	32	36	32	40	48	32	32	32	32	32	32	32	32	32	32	32
Gear on A	80	80	80	80	72	80	72	80	80	80	80	80	80	80	80	80	80
*Gross Product Per Hour	1200	1028	960	900	864	800	771	720	655	600	554	514	450	400	360	327	300
Time in Seconds To Make One Piece	3	3½	3¾	4	4⅛	4½	4⅔	5	5½	6	6½	7	8	9	10	11	12

SPINDLE SPEEDS

3000	2460	2110	1790	1530	1250	1090	895	765	650	560	455
1070	880	750	640	545	445	390	320	275	230	200	160
645	525	450	385	330	270	235	190	165	140	120	100

Rev. at Max. Speed to Feed Stock or Index Turret—½ Second

| 25 | 21 | 18 | 15 | 13 | 10 | 9 | 8 | 7 | 6 | 5 | 4 |

Rev. of Spindle at Max. Speed to Make One Piece

	25	21	18	15	13	10	9	8	7	6	5
150	123	105	90	77	63	54	45	38	33	28	23
175	144	123	104	89	73	64	52	45	38	33	27
188	154	132	112	96	78	68	56	48	41	35	28
200	164	141	119	102	83	73	60	51	43	37	30
208	171	146	124	106	87	76	62	53	45	39	32
225	185	158	134	115	94	82	67	57	49	42	34
233	191	164	139	119	97	85	70	60	51	44	35
250	205	176	149	128	104	91	74	64	54	47	38
275	226	193	164	140	115	100	82	70	60	51	42
300	246	211	179	153	125	109	90	77	65	56	45
325	267	229	194	166	135	118	97	83	70	61	49
350	287	246	209	179	146	127	104	89	76	65	53
400	328	281	239	204	167	145	119	102	87	75	61
450	369	316	269	230	188	163	134	115	98	84	68
500	410	352	298	255	208	182	149	128	108	93	76
550	451	387	328	281	229	200	164	140	119	103	83
600	492	422	358	306	250	218	179	153	130	112	91

650	533	457	388	332	271	236	194	166	141	121	99	13	277	80	32	36	78	4
700	574	492	418	357	292	254	209	179	152	131	106	14	257	80	32	36	84	4
750	615	527	448	383	313	272	224	191	163	140	114	15	240	60	60	80	80	4
875	718	615	522	446	365	318	261	223	190	163	133	17½	205	60	60	72	84	3
1000	820	703	597	510	417	363	298	255	217	187	152	20	180	60	60	54	72	3
1125	923	791	671	574	469	409	336	287	244	210	171	22½	160	60	60	48	72	3
1250	1025	879	746	638	521	454	373	319	271	234	190	25	144	60	60	48	80	3
1375	1128	967	821	701	573	499	410	351	298	257	208	27½	131	60	60	42	77	3
1500	1230	1055	895	765	625	545	448	383	325	280	227	30	120	40	80	60	60	3
1750	1435	1231	1044	893	729	636	522	447	379	327	265	35	103	40	80	60	70	3
2000	1640	1406	1194	1020	834	726	597	510	434	374	303	40	90	40	80	54	72	3
2250	1845	1582	1343	1148	938	817	671	574	488	420	341	45	80	40	80	48	72	3
2500	2050	1758	1492	1275	1042	908	746	638	542	467	379	50	72	40	80	48	80	3
2750	2255	1934	1641	1403	1146	999	821	702	596	514	417	55	65	40	80	42	77	3
3000	2460	2110	1790	1530	1250	1090	895	766	650	560	455	60	60	40	80	40	80	3
3375	2768	2373	2014	1721	1407	1226	1007	861	732	630	512	67½	53	36	72	40	90	3
3750	3075	2637	2238	1913	1563	1362	1119	957	813	701	569	75	48	36	77	40	90	3
4125	3383	2901	2462	2104	1719	1498	1231	1053	894	771	625	82½	44	36	90	35	90	3
4500	3690	3164	2686	2295	1876	1634	1343	1148	976	841	682	90	40	36	84	35	90	3
4875	3998	3428	2909	2486	2032	1771	1455	1244	1057	911	739	97½	37	32	78	36	96	3
5250	4305	3692	3133	2678	2188	1907	1567	1340	1138	981	796	105	34	24	80	40	84	3
5625	4613	3956	3357	2869	2345	2043	1679	1436	1220	1051	853	112½	32	32	90	36	96	3
6000	4920	4219	3581	3060	2501	2179	1790	1531	1301	1121	910	120	30	24	80	40	96	3
6750	5535	4747	4028	3443	2813	2452	2014	1723	1463	1261	1023	135	27	32	72	24	96	3
7500	6150	5274	4476	3825	3126	2724	2238	1914	1626	1401	1137	150	24	24	80	32	96	3
8250	6765	5801	4924	4208	3439	2996	2462	2105	1789	1541	1251	165	22	24	88	32	96	3
9000	7380	6329	5371	4590	3751	3269	2686	2297	1951	1681	1364	180	20	22	88	32	96	3
9750	7995	6856	5819	4973	4064	3541	2909	2488	2114	1821	1478	195	18	24	78	24	96	3
10500	8610	7384	6266	5355	4376	3814	3133	2680	2276	1961	1592	210	17	24	84	24	96	3
11250	9225	7911	6714	5738	4689	4086	3357	2871	2439	2102	1706	225	16	24	90	24	96	3
12000	9840	8438	7162	6120	5002	4358	3581	3062	2602	2242	1819	240	15	24	88	22	96	3

*Net will vary with factory conditions and the character of the work.

When threading to a shoulder, the necessary relative position of clutch teeth at time of reversing is obtained by using table values. This condition exists when a whole number is obtained by multiplying "time in seconds" by 48.

$$15 = \frac{\text{1st gear on B}}{\text{gear on A}} \times \frac{\text{gear on C}}{\text{2nd gear on B}} = \text{"time in seconds."}$$

Care should be taken to select combinations that will mesh but not interfere.

165

ing "time in seconds" by 48. Time in seconds is obtained as followed:

$$\text{Time in seconds} = \frac{\text{1st gear on } B}{\text{gear on } A} \times \frac{\text{gear on } C}{\text{2nd gear on } B}$$

Care should be taken to select combinations that will mesh but will not interfere.

The positions of the tools in the turret and on the cross slide for machining the workpiece shown in Fig. 6-15 are diagrammed in Fig. 6-18. The turret stop is used on this job.

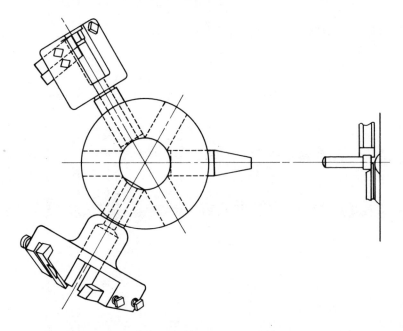

Fig. 6-18. Tool positions for machining the workpiece.

To estimate the spindle revolutions for the *balance turning tool*, several factors must be considered. Length turned is 1.500 inches, less 0.010 inch that is left under the head for the forming tool. Usually, from 0.010 inch to 0.015 inch is added to the length for all turret tools, with the exception of taps and dies. This extra length is required to permit the tools to approach the work at required feed and not "jab in" at the beginning of the cut; thus, (1.500 inch − 0.010 inch) + 0.010 inch = 1.500 inches as the length to be turned, or the throw of the cam lobe. The feed for the balance turning tool depends on the depth of cut (see Table 6-1), which in this instance is:

$$\text{Depth of cut} = \frac{\text{diameter of stock } - \text{ finished diameter}}{2}$$

$$= \frac{1.000 - 0.625}{2} = \frac{0.375}{2} = 0.1875$$

Therefore, the feed selected from the table (see Table 6-1) is 0.010 inch per spindle revolution, which is the feed for 3/16-inch depth of cut. Dividing 0.010 inch into 1.5 inches (length of cut) gives 150 revolutions (1.5 ÷ 0.010 = 150) of the spindle required for the operation.

For the *box tool*, the turned length is the same as for the balance turning tool. A feed of 0.010 inch per revolution (see Table 6-1) is used for box tools when finish turning, removing a depth of 0.005 inch, which was the amount left by the balance turning tool. Therefore, this operation requires 150 spindle revolutions (1.5 ÷ 0.010 = 150).

The *form tool* must advance on the workpiece a distance equal to one-half the difference in the diameter of the stock and the finished diameter, or:

$$\text{Advance} = \frac{1.000 \text{ inch } - 0.625 \text{ inch}}{2}$$

$$= \frac{0.375 \text{ inch}}{2} = 0.187 \text{ inch}$$

If a distance of approximately 0.005 inch is allowed for approaching the work with the form tools, then:

0.187 inch + 0.005 inch = 0.192 inch throw

The cutoff tool will have considerable work to do after the forming tool had completed its portion of the work (see Fig 6-15); therefore, the cutoff tool should be started first. Thus, the forming

167

tool shaves material off the sides of the head only, until it approaches full depth. Therefore, it is possible to feed inward until within about 0.010 inch of full depth.

From the table (see Table 6-1), a feed of 0.003 inch per revolution is selected for the first 0.182 inches, and 0.001 inch per revolution is selected for the final 0.010 inch of throw:

$$0.182 \text{ inch} \div 0.003 \text{ inch} = 61 \text{ spindle revolutions}$$
$$0.010 \text{ inch} \div 0.001 \text{ inch} = 10 \text{ spindle revolutions}$$

If 8 spindle revolutions are added to include "dwell" for the last 0.010 inch of travel, a total of 78 spindle revolutions are required for the form tool

The *cutoff tool* must advance on the workpiece a distance equal to one-half the diameter of stock plus a distance of 0.004 inch to 0.008 inch for approach to the work, and a distance large enough to permit the heel of the tool to pass the center and trim the end of the bar, leaving no material. The angles and thicknesses for circular cutoff tools are given in Table 6-3. For example, a cutoff tool for 1-inch brass stock has an angle 0.059 inch in depth. Usually, an allowance of 0.003 inch to 0.005 inch is made for the heel to pass the center point when cutting off a solid bar. The throw of the cutoff tool in this instance is as follows:

One-half diameter of stock = 0.500 inch (1.000 ÷ 2)
Approach to work   = 0.010 inch
Depth of angle on tool   = 0.059 inch
Heel of tool past center   = 0.005 inch
Total throw for cutoff tool = 0.574 inch

The thickness of the cutoff for stock 1 inch in diameter is 0.140 inch (see Table 6-3), or approximately ⅛ inch, and a circular tool of that thickness, or width, is fed 0.0035 inch per spindle revolution (see Table 6-1). Therefore, 0.574 inch ÷ 0.0035 inch = 164 spindle revolutions required for the cutoff tool.

As 164 spindle revolutions are required for the cutoff tool and only 78 revolutions are required for the forming tool, the spindle revolutions for forming and cutoff can be overlapped. Then, as the work progresses, the spindle revolutions required for all opera-

Table 6-3. Angles and Thicknesses for Circular Cutoff Tools

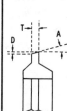

A is 23° when cutting brass, aluminum, copper, silver and zinc.

A is 15° when cutting steel, iron, bronze, and nickel.

Least thickness used when cutting off into tapped holes is the
lead of two and one-half threads plus 0.010".

Least thickness used when cutting off into reamed holes smaller
than ⅛" diameter is 0.040".

Thickness used when cutting off tubing is two-thirds T as given
below for corresponding diameters of stock.

Thickness used when angles or radii start from outside diameter
of tool is governed by varying conditions and determined
accordingly.

Diameter of Stock		T Thickness	D Depth of Angle	
Fractional	Decimal		for Brass	for Steel
¹⁄₁₆	0.06250	0.020	0.0085	0.0055
³⁄₃₂	0.09375	0.030	0.013	0.008
⅛	0.12500	0.040	0.017	0.011
³⁄₁₆	0.18750	0.050	0.0215	0.0135
¼	0.25000	0.060	0.0255	0.016
⁵⁄₁₆	0.31250	0.070	0.030	0.019
⅜	0.37500	0.080	0.034	0.021
⁷⁄₁₆	0.43750	0.090	0.038	0.024
½ to ⁹⁄₁₆	0.5000 to 0.5625	0.100	0.042	0.027
⅝ to ¾	0.6250 to 0.7500	0.120	0.051	0.032
¹³⁄₁₆ to 1	0.8125 to 1.000	0.140	0.059	0.038
1¹⁄₁₆ to 1⁵⁄₁₆	1.0625 to 1.3125	0.160	0.068	0.043
1⅜ to 1⅞	1.3750 to 1.8750	0.190	0.081	0.051
2 to 2½	2.0000 to 2.5000	0.220	0.093	0.059

tions that actually take time can be entered in the cam design work
sheet (see Fig. 6-16) in the second set of columns, eliminating those
that overlap.

Spindle revolutions are also required for *indexing the turret* and
feeding stock. To find the spindle revolutions for these operations,
take a trial total of the revolutions that remain after the overlapped
operations are removed.

Rough turning requires 150 revolutions, finish turning requires
150 revolutions, and cutting off requires 164 revolutions. These
operations total 475 revolutions to give an approximate idea of the
time required for the job.

On referring to the table for laying out cams (see Table 6-2), it
can be observed that the total is within the limit of 20 seconds, the
point at which the cam surfaces normally begin to govern the time

required for feeding stock or indexing the turret on this machine. The revolutions required for these movements are calculated on the basis of time required for the mechanism to operate plus a few extra revolutions to allow time for setting the trip dogs. It sometimes seems, from the time trial of revolutions, that only actual time is required for feeding of stock and indexing the turret; but on proceeding with entries in the work sheet, it is found that the cam surface governs the time required for these movements.

To feed stock or index the turret, ½ second is required (see Table 6-2); thus there are 25 spindle revolutions at 3000 r/min. If an additional 5 revolutions are added, as mentioned, 30 revolutions should be allowed for feeding stock, and another 30 revolutions for each time that the turret is indexed.

Tool clearance is also important and must be considered. It is desirable that the cross slide tools begin to cut as soon as possible after the turret tools have completed their work. In some instances, there is no interference; but on work that is turned to within a short distance of the chuck, ample time must be allowed for the turret tool to move backward, so that it will clear the advancing cross slide tools.

In this example, the cross slide tools must clear a No. 22B box tool. By referring to Table 6-4, we find that 7 hundredths of cam circumference for the front cross slide tool and 4 hundredths for the back cross slide tool are required. As there is ample time for the forming operation while the workpiece is being cut off, the clearance of 4 hundredths of cam circumference for the back cross slide tool is the only extra time required to be added to the actual time for the job; the balance of the clearance, 3 hundredths, required for the front cross slide clearance can be overlapped into the time required for cutting off.

After the estimate of the number of spindle revolutions required for each tool and operation has been completed and entered on the work sheet, the column headed "Spindle Revolutions After Deduction for Operations Overlapped" can be added. The sum, in this instance, is 554 spindle revolutions.

If 4 hundredths of the cam circumference must be devoted to clearance, as mentioned, the balance, or 96 hundredths, represents the total number of spindle revolutions, 554, for the operations and idle movements. Dividing the total by 96, and multiplying by 100

Table 6-4. Approximate Clearance Between Turret Tools and Cross Slide Circular Tools (in Hundredths of Cam Surface)

NO. ON TOOL	TURRET TOOLS	CLEARANCE									
		Front Cross Slide Tool					Back Cross Slide Tool				
		No. 00	No. 0	No. 2	4	6	No. 00	No. 0	No. 2	4	6
20-22	Balance Turning Tool		7	7	6	0		7	7	6	6
24-26	Balance Turning Tool				6					6	6
00A	Balance Turning Tool	6					6				
00B	Balance Turning Tool	6					6				
00K-20K-22	Box Tool	8	7	7	7		6				
24-26	Box Tool				7	7	6	7	6		
00L-20L	Box Tool	6	6				6	7			
00BM-20B-22B	Box Tool	8	7	7			7	5			
00CA-20CM	Box Tool	8	7				5	6	4		
00D-20H	Box Tool	5	7				6	6			
00EA-00FA-20E-22E	Box Tool	8	7	7			6	6	6		
00-00CA-11-22	Centering and Facing Tool	8	7	7			4	3	3		
00E-20-22	Die Holder	7	6	6			6	6	6		
24-26	Die Holder										
00-00BA-20-22	Knee Tool	6	5	5	6	6	4	4	4	6	6
00-20-22	Knurl Holder	7	6	6	6	6	6	6	7	5	5
24-26	Knurl Holder									5	6
00B-20B-20D-22B-22D	Pointing Tool	7	7	7	6	6	5	6	6		
00C	Pointing Tool	8					6				
24-26	Pointing Tool				6	6				6	6
00D	Pointing Tool	5			3	3	7				
24B-26B	Turret Tool Post	5									

NOTE—For No. 00 Size Machines—

Double the time given in table for No. 00 Size Machines, if a 3 second job.

On a 4 to 5 second job, add from 4 to 5 hundredths of cam surface to figures given in table for these machines.

171

(554 ÷ 96 × 100) gives a total of 576, which is the total estimated number of revolutions required to finish the workpiece.

Selecting actual spindle revolutions—It can be noted that change gears are not provided for exactly 576 revolutions at a speed of 3000 r/min (see Table 6-2), so it is necessary to select either 600 revolutions or 550 revolutions, as these are the nearest figures for which change gears are available. If the number of spindle revolutions is *larger* that the estimated number, *add* the extra revolutions to one or more of the cutting operations; if the number is *smaller* than the estimated number of revolutions, *subtract* from the cutting operations. Care should be taken not to attempt to subtract too many revolutions because the feeds can be increased beyond the limit for obtaining the desired finish.

In this example, 600 revolutions are selected from the table, which indicates that extra revolutions must be added to the total estimated revolutions (576). When the number selected is larger than the estimated total, the additional revolutions required to bring the total to 600 revolutions can be calculated as follows:

$$600 \times 0.96 = 576 \text{ revolutions}$$

Subtract   554 revolutions (96% of estimated total spindle revolutions)

          22 revolutions to be added

In this example, the 22 extra revolutions are added to the cutoff time because that is the operation to which they can be added most advantageously. It can be noted in the table (see Table 6-2) that 600 revolutions represent a 12-second operation, and the gross production per hour is 300 pieces of work.

Determining the hundredths of cam surface—The hundredths portions of cam surface required for each operation and idle movement can be found by dividing the number of revolutions required by the total number of revolutions required, using the nearest hundredth. The total of the hundredths portions must equal 100, or the full circumference of the cam.

Drawing the Cams

After the calculations have been completed and the portion of the cam surface required for each operation has been determined, the cam can be drawn. The usual simple and inexpensive drawing equipment is needed; in addition, a cardboard template for quickly marking the hundredths divisions on the cam circle and cam templates for drawing the quick rise and fall of the cam lever rolls when either bringing tools into position or dropping backward are needed. Many firms that have a number of machines use cam sheets printed with a cam circle divided into hundredths. The sheet also has spaces available for listing the tools used, order of operations, and other essential information that is needed. The cam design work sheet shown in Fig. 6-16 is used for the workpiece shown in Fig. 6-15.

The important steps in drawing the cam (Fig. 6-19) are as follows:

1. Draw a long horizontal line to represent the center line near the top of the sheet.
2. On the left-hand side, draw a vertical line to indicate the position of the face of the chuck.
3. Draw the workpiece and the cutoff and form tools as near full scale as possible.

Establishing position of the turret slide—The least distance of the turret from the face of the chuck on the No. 2 high-speed

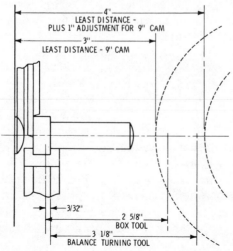

4"
LEAST DISTANCE -
PLUS 1" ADJUSTMENT FOR 9" CAM
3"
LEAST DISTANCE - 9" CAM

3/32"
2 5/8"
BOX TOOL
3 1/8"
BALANCE TURNING TOOL

Fig. 6-19. Layout diagram for the cam.

173

machine (36 spindle speeds) is 3 inches, as indicated in Table 6-5. The turret is in this position when the lead lever roll is on the circumference of a 9-inch cam. A screw adjustment on the turret slide permits setting the turret back a distance of 1 inch, if required, so that the least distance from the turret to the face of the chuck is then between 3 and 4 inches, depending on the distance the slide is set back. Then, if enough space still cannot be obtained, the cam must be reduced to a smaller diameter on some or all of the lobes.

To determine the distance at which to set the turret slide and the amount to reduce the lead cam lobes, refer to the dimensions of stock tools (Table 6-6). The table gives the length of the body and shank on both the No. 22 balance turning tool and the No. 22B box tool. From the table, the length of the body of the balance turning is 3 ⅛ (3.125) inches.

Lay out the distance (3.125 inches) on the scale drawing (see Fig. 6-19), beginning at the point where the tool finishes its cut at the shoulder of the piece and extending backward toward the turret position. Repeat this procedure with the box tool. Both these points are established beyond the position of the turret when it is at its shortest distance from the chuck.

If the full adjustment of the turret slide is used, there will be barely enough space for the balance turning tool. On a 9-inch cam blank, the shortest distance from the turret to the chuck is 3 inches; then the additional distance desired for the balance turning tool can be obtained by adjusting the turret slide backward 0.5 inch, plus a 0.5-inch reduction on the cam. *Note:* With a 9-inch cam, a length of 2 inches can be turned on the workpiece; with an 8-inch cam, a 1.5-inch length can be turned. If none of the throws on the cam exceed 1.5 inches and reductions are not required, an 8-inch cam can be used. By using special long turret travel parts, 3 inches can be turned with a 9-inch cam.

At all times it is desirable to have the tools set back into the turret, as close to the turret as possible. If the shanks extend through the holes into the center of the turret, care must be exercised to see that two tools in succession do not extend through, as the shanks can interfere with the operation of the turret on the machine.

Drawing the cam lobes—First, draw a 9-inch circle and divide

174

the circumference into hundredths, using a template. From the cam design sheet (see Fig. 6-16), indicate the hundredths divisions or graduations on the cam circle as to where operations are to begin and end for the turret slide or lead cam.

The hundredth division or the graduation line at the top, directly above the hole for the locating pin, is the "zero" line. The stop for the stock is in position for the first five divisions of the circle, and a "dwell" is drawn on the cam surface. If this is drawn to the full height of the cam, a high narrow lobe results; therefore, it is better to reduce the cam at a point 1.25 inches from full height. A stop of proper length can be used in the turret to compensate for this reduction.

From the 5th to the 10th division marks, the turret is double indexed and dropped backward to the required position for beginning the first turning operation. On the 10th division radius line, measure downward and locate a point 1.5 inches from the outside diameter, plus a 0.5-inch reduction—this represents the beginning of the balance turning tool lobe.

Locate the cam template at the center of the drawing with an ordinary pin through the hole in the template (Fig. 6-20). Place the

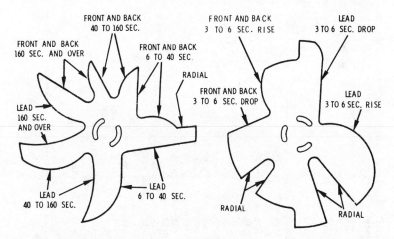

Fig. 6-20. Two cam templates for Brown & Sharpe No. 2 machines. One is used for those jobs that are within the range of "6 to 240 seconds" per piece (left), and the other is a high-speed template for jobs within the range of "3 to 6 seconds" per piece (right).

175

Table 6-5. Machine Capacities and Speeds (High 6-Speed and Screw Threading Machines)

After Serial Nos. Indicated

	Nos. OOH.S. (3 Spds.), OOH.S. (36 Spds.), OOG H.S. (Beginning Serial No. 12852)	Nos. OH.S. (3 Spds.), OH.S. (36 Spds.), OG H.S. (Beginning Serial No. 8021)	Nos. 2H.S. (3 Spds.), 2H.S. (36 Spds.), 2G H.S. (Beginning Serial No. 7352)	Nos. OOH.S.. OOG H.S. Screw Thrd. (Begin Serial No. 13432)
Diameter of hole through regular feed tube	25/64″	21/32″	1 1/8″	25/64″
Largest stock taken in regular feeding fingers ... Round	3/8″	5/8″	1″	3/8″
Hex.	5/16″	1/2″	7/8″	5/16″
Square	1/4″	7/16″	11/16″	1/4″
NOTE—For special arrangements for greater capacities, see below.				
Greatest length that can be turned at one movement	3/4″	1 1/4″	2″	
Greatest length that can be fed at one movement	1″	2″	2 1/2″	1″
Number of spindle speeds	3, 36, 36	3, 36, 36	3, 36, 36	10
Fastest and slowest spindle speeds r/min	6000, 6000, 6000 / 1200, 200	4220, 4220, 4150 / 850, 150, 155	3000, 3000, 3000 / 645, 100	5000 / 1212
Change gears give one revolution Fastest	3/4 sec.	1 2/3 sec.	3 sec.	3/4 sec.
of cams Slowest	45 1/2 sec.	176 1/2 sec.	240 sec.	20 sec.
Actual time allowed to feed stock	1/4 sec.	1/3 sec.	1/2 sec.	1/4 sec.
Actual time allowed to index turret	1/4 sec.	1/3 sec.	1/2 sec.	
Number of holes in turret	6	6	6	
Diameter of holes in turret	5/8″	3/4″	1″	
Diameter of turret	3 3/8″	4 1/16″	5″	

	Nos. OOH.S. (3 Spds.), OOH.S. (36 Spds.), OOG H.S. (Beginning Serial No. 12852)	Nos. OH.S. (3 Spds.), OH.S. (36 Spds.), OG H.S. (Beginning Serial No. 8021)	Nos. 2H.S. (3 Spds.), 2H.S. (36 Spds.), 2G H.S. (Beginning Serial No. 7352)	Nos. OOH.S., OOG H.S. Screw Thrd. (Begin Serial No. 13432)
Greatest distance tools can project from turret	2¼″	3¼″	3⅞″	
Greatest diameter of tool turret will swing	1⅝″	2⅝″	2⅞″	
Greatest distance between turret and chuck	2 15/16″	5⅛″	6¾″	3″
Least distance between turret and chuck	1⅞″	2½″	3″	3/16″
Center of holes in turret to side of turret slide	⅞″	1⅜″	1½″	
Screw adjustments of turret slide		¾″	1″	
Rack adjustment of turret slide (one tooth)	Approx. 3/16″			
Top of cross slide to center of spindle	1″	1 5/16″	1 7/16″	1″
Number of Changes of Speeds				20
Die Spindle Spindle Speeds, r/min. Fastest				7000
Slowest				1454
Number of Cutting Speeds				20
Threading Cutting Speeds, r/min. Fastest				2000
Slowest				242
Movement of cross slide	1″	1¼″	1¾″	1″
Distance from center of spindle to floor	46″	46″	46″	46″
H.P. required at maximum capacity	2	3	5	2¼″
Floor space, length	48″ 48″ 53″	62″ 62″ 66″	68″ 68″ 71″	48″ 60″
Floor space, width	27″ 27″ 43″	30″ 30″ 47″	34″ 34″ 48″	27″ 43″
Net weight, about, in lbs.	1400 1600 1600	1850 2075 2075	2500 2875 2975	1600 1700

Table 6-6. Principal Dimensions of Stock Tools

(No. 2 Machines)

NOTE: If two capacities are given for one tool, the capacity giving greater length and smaller diameter can be obtained only by allowing the work to pass through the hole in the shank.

Tool	Length Body	Length shank	Capacity Dia.	Capacity Length	Hole Shank	Distance from Cutting Edge to Front End of Body
No. 22DA Adjustable Guide—Center to tongue to front of guide. 3 5/8"						
No. 22B Adjustable Guide, left-hand	3 5/16	1 3/4	Angle 30° inc.		5/8	
No. 22 Angular Cutting-Off Tool	1 3/4	2 3/8	Angle 30° inc.		5/8	
No. 22A Back Rest for Turret	1 7/8	2 1/4	50° to 80° inc.		1/2	
No. 22C Back Rest for Swing Tool	3 1/8	1 3/4	3/32 to 1/8		1/2	
No. 22 Balance Turning Tool	3	1 3/4	1/4 to 1	2 5/8	1/2	
No. 22C Balance Turning Tool	2 5/8	2 1/4	1/4 to 1	2 5/8	1/2	
No. 22A Box Tool	2 5/8	2 1/4	5/8	2	1/2	
No. 22B Box Tool			3/16 to 5/8	2	1/2	
No. 22G Box Tool, three blades, adjustable; greatest distance between blades 1 7/8", least 1/16	3 1/8	2	7/8 to 5/8	2 3/8	1/2	3/32
No. 22E Box Tool, roller back rest	3 1/8	2	5/8 to 3/4	3	11/16	1/8
No. 22FA Box Tool, roller back rest, left-hand	3 1/8	2	5/16 to 5/8	2 5/8	11/16	1/8
No. 22D Centering and Facing Tool, center drill 5/8"	1 3/4	2 3/4	5/16 to 7/8	3	5/8	1/16
No. 22CA Centering and Facing Tool, center drill 5/8"	1 3/4	2 3/4		2 3/8	5/8	1/16
No. 22 Combination Drilling and Tapping Attachment						
Drill	1 1/2					
Tap	1 9/16		Brass 3/8–24 Steel 1/4–32			
No. 22 Cross Drilling Attachment, center of tongue to end of drill spindle 2 15/16"						
No. 22B Die Holder, releasing, extreme pull out 1/8"	2 3/8	2	1/4	2 3/8	1/4	
No. 22 Die Holder, extreme pull out 1/2"	2 3/8	2	1/2	1 3/4	3/8	

No. 733B	Die Holder, Acorn, releasing	3 1/4	2	3/8–1/2 1/2–5/8	3 1/8 2 1/8		
No. 22	Drill Holder	1 1/2	1 3/4			5/8	
No. 22	Drilling Attachment	1 1/2		3/8		1/4	
No. 22E	Fixed Guide; center of tongue to front of guide 3 1/2"						
No. 22D	Fixed Guide; center of tongue to front of guide 3 5/8" (For L.H. swing tools)						
No.22	Floating Holder	1 9/16	1 3/4			11/16	
No. 22A No. 22C	Hollow Mill, Plain, held in Floating Holder	2 5/8	1 3/4	7/16 11/16			
No. 22	Knee Tool	3	2	21/32 1 7/16	3 1 11/16	11/16	1/8
No. 22B	Knee Tool	3		21/32 17/16	3 1 1/4	11/16	1/8
No. 22	Knurl Holder for Turret	2 3/8	2 3/16	3/16 to 15/16 3/16 to 1 1/8	1 7/8 1/2	1/2	
No. 22AA	Knurl Holder, side; center tool to center knurl 1 7/32"			width 1/4			
No. 22BA	Knurl Holder, top, back; center of circular tool to center of knurl 2 3/16"			1 1/8 width 1/4			
No. 22C	Knurl Holder, bottom; center of circular tool to center of knurl, 2 3/8"			1 1/4 width 1/4			
No. 22A	Pointing Tool Holder, for circular tools	2 3/8	1 25/32				
Nos. 22B and 22D	Pointing Tool, box	2	2 1/2	7/8		5/8	1/2
	Stock Stops, length 2 1/2", 2 7/8", 3 1/4", 3 5/8", 4"						
No. 22M	Recessing Tool, diameter of shank 1"	2	1 5/8				
No. 22BA	Swing Tool, actual projection of body 2 3/4"	2 17/32	2	21/32 1	3 1 3/4	11/16	
No. 22CA	Swing Tool, actual projection of body 2 1/16"	1 7/8	2	21/32 1	3 1 1/8	11/16	
No. 22HA	Swing Tool, recessing; swing from center 1/8"	2 1/4	2			5/8	
No. 22KA	Swing Tool, knurling or thread rolling	2 3/16	2 1/2	21/32 1 3/8	3 1 1/4	11/16	1/4
No. 22LA	Swing Tool, actual projection of body 2 3/4"	2 15/32	2 1/2	21/32 1	3 1 3/4	11/16	
No. 22	Tap Holder; extreme pull out 1/2"	1 5/8	2				
No. 22B	Tap Holder, releasing; extreme pull out 1/8"	2 1/16	2				
No. 22	Tapping Attachment; extreme pull out 5/16"	1 9/16		Brass 3/8–24 Steel 1/4–32			

arm that has a drop marked "6 to 40 seconds" so that the edge is on the 5th division line, and follow the edge with a pencil. Set the compass to the radius of the cam roll ($\frac{5}{16}$ or 0.3125 inch), and connect the line of drop with the point on the 10th division radius line where the lobe for the balance turning tool begins. The space for indexing allows some clearance after the roll has ended contact with the line of drop and before it starts on the balance turning lobe. This clearance should be provided for because it is not always safe to drop from one lobe directly to the starting point of the next lobe.

How to develop a cam lobe curve—The balance turning tool is in operation from the *10th to the 35th division line*. To develop a uniform curve with constant rise, divide the throw into any number of equal parts, and then divide the arc of the cam circumference that it occupies into the same number of equal parts. Swing the arc at the division points of the throw, and draw the radii to intersect these arcs from each division point on the circumference of the cam. Use a scroll to draw a spiral curve through the points of intersection to give the outline of the cam lobe.

The indexing space from the *35th to the 40th division line* and the lobe for the box tool from the *40th to the 65th division line* are drawn in exactly the same manner as has been described previously; then the outline drops back to the full depth of the throw, withdrawing the turret slide while the cross slides are in operation. The turret slide advances again, according to the rise of the template, in time for the stock to stop again at the correct distance (zero).

The cross slide tools begin to cut as soon as possible after the box tool is finished. The diameter of the cross slide cams is 7 inches; use the same center, but indicate the outline of the front cross slide cam by dots and the back cross slide cam by alternate dots and dashes.

As indicated on the work sheet, a 4 hundredths space is required for clearance between the box tool and the back cross slide tool, so the cross slide tool begins at the *69th division line and ends at the "zero" division line*, as the operation requires 31 hundredths of the cam surface.

The starting point of the throw is measured from outside the cam, and the rise from the full depth of the cam to the starting point of the lobe is made by the cross slide curve on the cam template. The lobe curve is developed in the same manner as described previously. There is no difference in drawing the lobe for the front slide or forming tool, except that more clearance is necessary, and during the last hundredth space (from 84½ to 85), the curve is on the true circumference of the cam so that the tool dwells for several revolutions before dropping back.

Marking the drawing—It is a desirable practice to place all essential information on a drawing for the purpose of making a record and to make the job easier for the person making the cams. Note that a dimensioned drawing can be made in the blank space on the cam design work sheet. At the top of the drawing, the distance to adjust the turret from the chuck (with the cam roll on top of the stop lobe) can be indicated for the convenience of the operator setting up the job.

It is important for the toolmaker to understand the design of such special-purpose cams. However, industry is adapting new technology to the task of designing such cams. A computer program for the type of special-purpose cam needed is prepared. The necessary information is fed into a computer connected to a plotter. The cam is drawn by the plotter. After the program is prepared, the process takes less than 15 minutes total time. Manual methods usually took from 25 to 30 hours for the same task. With such a saving of time, it should be expected that many more tasks will be assigned to the computer.

HOW TO MACHINE CAMS

Any good homogeneous sheet steel of the required grade of steel is suitable for cams. When they are hardened properly, cams from such stock can withstand the demands of all ordinary requirements.

Generally, it is more convenient to make the cams from stock blanks, as they can be turned to correct size, drilled for the shaft and locating pin, and the cam surface graduated into hundredths division lines to facilitate laying out the outline. If the blanks are made in the shop, the holes must be of the correct size so that there will be no play of the cam on the shaft.

Transferring the cam outline—To get successful results, the cam outline as plotted on the drawing should be carefully and closely reproduced on the blank. A quick and handy method of transferring the outline is to place a piece of carbon paper and a piece of stiff plain paper beneath the drawing and trace the outline of the drawing with a dull steel point, which makes a carbon outline on the plain paper. This can be cut out and used as a template for scribing the outline on the cam blank.

Machine the cam outline—After the outline is laid out on the cam blank, it is necessary first to remove the excess metal close to the outline between the lobes. This metal can be removed by drilling a row of holes near the outline, breaking off the excess metal; a band saw or similar means can be used.

Three methods of finishing the curves of cam lobes are in use. The simplest method is filing the excess metal down to the outline. This is a slow procedure and requires considerable care to obtain a smooth curve. Cam lobes produced by this method can be satisfactory if the limits of accuracy are not too close, but on finer classes of work the outline of the cam lobe must be an accurately increased curve for the tool to be fed uniformly throughout the entire throw of the lobe. An extremely accurate curve can be produced on the milling machine geared for the correct lead.

A second method of finishing the cam lobes after they have been drilled and broken off to the approximate outline requires the use of a universal or plain milling machine equipped with an index head (Fig. 6.21). The cam blank is mounted on the index head spindle, which should be parallel to and directly beneath the machine spindle. An end mill having the same diameter as the cam roll is used in the setup. As shown in the setup, the end mill is set at the top of the lobe and then indexed a small amount. Then the table is fed a vertical distance corresponding to the rise of the lobe in the distance that the blank is indexed. These operations are continued until the curve of the lobe is completed. The surface that is produced may require a slight amount of filing, depending on the fineness of the division selected for indexing the blank and the finish desired.

The third method of finishing the cam lobes is nearly as simple and is much preferred to the other two methods, as it produces a true uniform curve without recourse to filing. Either a plain or a

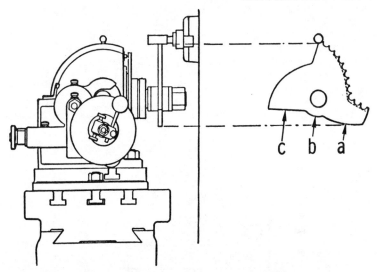

Fig. 6-21. Milling machine setup for finishing the cam lobes.

universal milling machine equipped with a spiral headstock and a vertical milling attachment can be used.

The spiral head is geared to the table feed screw, as in cutting spirals, and the cam blank is fastened to the end of the index spindle.

An end mill is used in the vertical milling attachment, which is set to mill the periphery of the cam at right angles to its sides. In other words, the axes of the spiral head spindle and the attachment spindle must always be parallel in this method of milling cams. The cutting is done by the teeth on the periphery of the end mill.

To illustrate the basic principle of this method, the spiral head is elevated to 90°, or at right angles, to the surface of the table and is geared for any given lead. As the table advances and the blank is turned, the distance between the axes of the index spindle and the attachment becomes less, which indicates that the cut becomes deeper and the radius of the cam is shortened, producing a spiral lobe having a lead that is the same as that for which the machine is geared (Fig. 6-22).

If the same gearing is retained and the spiral head is set at "zero" reading, or parallel to the surface of the table, it is apparent that the axes of the index spindle and the attachment spindle are parallel

183

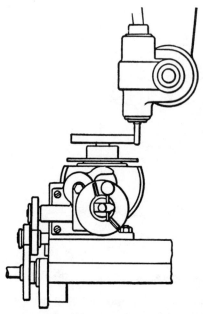

Fig. 6-22. Setup for machining a cam blank with the axis of the spiral head at 90° to the surface of the table.

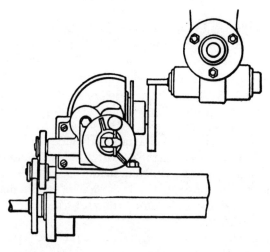

Fig. 6-23. Milling machine setup for finishing cam lobes in which the axis of the spiral head is parallel to the surface of the table.

(Fig. 6-23). Therefore, as the table advances and the blank is turned, the distance between the axes of the index spindle and the attachment spindle remains the same. As a result, the periphery of the milled blank is concentric, or the lead is "zero."

If the spiral head is elevated to any angle between 0° and 90° (Fig. 6-24), the amount of lead given the cam will be between the lead for which the machine is geared and "zero." Hence, cams with a large range of different leads can be obtained with a single set of change gears; the problem of milling the lobes of a cam is reduced to determining the proper angle to set the head to obtain a given lead. Manufacturers can provide a set of tables from which the proper angles and change gears can be selected.

SUMMARY

The cam is a device used to convert rotary motion into reciprocating motion. Cams can be generally divided into two classes: uniform motion and uniformly accelerated cams.

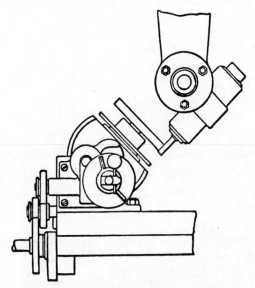

Fig. 6-24. Setup for machining a cam, with the spiral head at an oblique angle to the surface of the table.

The motion of a cam is transmitted to a follower, which remains in contact with the acting surface as the cam revolves. The uniform motion cam moves the follower at the same rate of speed throughout the stroke, from the beginning to the end. This means that if the movement is rapid, there is a distinct shock at the beginning and end of the stroke because the movement is accelerated from zero to the top speed of the uniform motion almost instantly and is stopped just as abruptly.

The uniformly accelerated motion cam is one in which the speed is slow at the beginning of the stroke, accelerated at a uniform rate until maximum speed is attained, and then uniformly retarded until the follower is stopped or nearly stopped when the reversal takes place.

REVIEW QUESTIONS

1. What is the main purpose of the cam?
2. What are uniform motion cams?
3. What is gravity motion?
4. Explain uniformly accelerated motion cams.
5. What type of cam is used in gasoline engines?
6. What are the classes of cams?
7. What are the commonly used followers?
8. What is the purpose of the displacement diagram?

CHAPTER 7

Dies and Diemaking

A die is a tool intended to *cut* or *shape* sheet metal or similar materials. Basically, the die proper consists of: (1) the die block and (2) the punch.

The *die block* is the part of the die that contains a hole that has the same outline as the piece that is to be punched. The *punch* fits into the hole in the die block; it is the part to which power is applied for performing the cutting operation. The hardened and tempered die block does the actual cutting. The *stripper plate* in the die block strips the stock from the punch on the releasing stroke. The *guide strip* merely guides the stock, and the *gage pin* is used to gage the location of the holes punched in the stock.

In some die designs, the parts of the punch are integral in one piece, but they are separate pieces united by suitable means in other designs. The *shank* attaches the punch to the ram of the press, and the *collar* of the *punching block* takes the thrust.

Fig. 7-1 shows typical sheet metal stampings produced for the electronics industry.

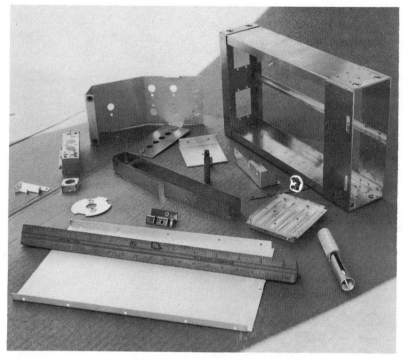

Fig. 7-1. Typical sheet metal stampings.

CUTTING OR PUNCHING DIES

Many types of dies are required to handle the large variety of cutting and shaping operations encountered in general practice. Dies can be classified with respect to the basic operation that they perform: (1) cutting dies, (2) shaping dies, (3) combining cutting and shaping dies.

The cutting dies are the simplest form of dies. The part that is cut out can have any shape, depending on the shape of the punch.

Plain Die

The plain, or blanking, die is the most frequently used type of die. Depending on the shape of the punch, the *blank* can have any shape. This die is used to punch a given shape from sheet metal in one continuous cutting operation (Fig. 7-2). The material can be either stock or a blank obtained from a preceding operation. Sometimes the blank punched from the stock is the part to be used; or it can be only waste material, and the punched stock is the portion that is to be used.

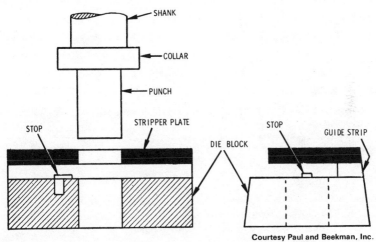

Courtesy Paul and Beekman, Inc.

Fig. 7-2. Diagram showing basic construction of a plain die.

A typical sequence of operations of a plain die in punching sheet metal is illustrated in Fig. 7-3. A strip of sheet metal or stock is placed between the die and the stripper plate; on the downward stroke, the punch "punches out" the blank from the stock, and the stock is held down by the stripper plate to prevent lifting on the upward stroke of the punch. If several punches are combined for blanking several pieces simultaneously, they are called *multiple dies.*

Self-Centering Die

A self-centering die is provided with a small conical point (Fig. 7-4) so that it does not require a gage pin for locating the holes punched in the stock. The self-centering die is used chiefly for punching holes. The stock must be center punched at the points

189

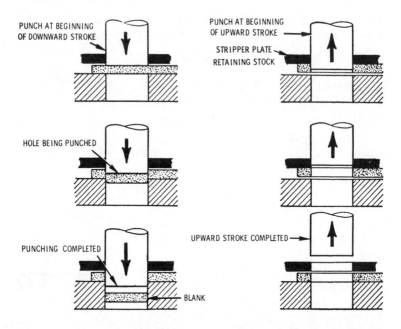

Fig. 7-3. Sequence of operations of a plain die in punching sheet metal, showing the downward (or punching) stroke (left) and the upward (or releasing) stroke (right).

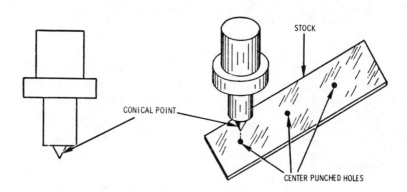

Fig. 7-4. A self-centering die (left), and view showing the die in correct position (right).

190

where the holes are required and then placed under the punch so that the conical point registers with the center punch mark on the stock.

SHAPING DIES

Shaping dies are used to change the shape of the stock; they do no cutting. The term "punch," strictly speaking, is a misnomer because it does not operate as a punch in shaping operations. The punch is used to press the stock against the die to bend the stock to the desired shape. Fig. 7-5 shows some sheet metal parts that

Fig. 7-5. Parts that require shaping.

require shaping. As will be noted, some of these are combined with cutting or punching. These may be separate operations or combined, depending upon the particular job.

Plain Bending Die

A typical die used to bend stock to the desired shape is the plain bending die, shown in Fig. 7-6. The contour of the punch and die determines the final shape of the stock.

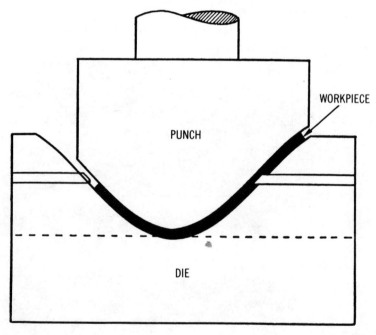

Fig. 7-6. A plain bending die.

The springiness of the stock or material is an important factor in this type of die operation. In bending stock to irregular shapes (Fig. 7-7), an allowance must be made for this factor in the shape of the die, so that the work is bent slightly beyond the required angle or shape. This allowance is necessary to compensate for the backward spring of the stock when it is released from the die.

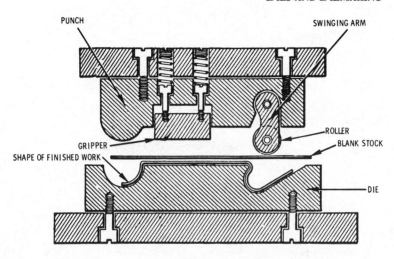

Fig. 7-7. A typical die for bending stock to an irregular shape.

Curling Die

A curling die is a special form of bending die in which part of the stock is bent to form circular beads around the edge of drawn cylindrical parts (Fig. 7-8). The progressive operations in curling a bead around a cylindrical vessel are shown in Fig. 7-9.

The curling punch forces the metal outward to increase the circumference at the edge of the metal. The metal will crack if it is overstretched. The stretching action of the metal increases as the diameter of the bead increases. Ordinarily, the maximum diameter of the bead that can be formed is $3/16$ inch for tinplate and sheet iron. An extra large bead can be obtained without cracking the metal by annealing the metal.

Wiring Die

This is a modification of the curling die in which provision is made for curling the edge of the workpiece over a wire. The basic diagram of a wiring die is shown in Fig. 7-10. The shape of the punch is the same as for the curling die, but the die proper (floating ring) is constructed to "float" on springs, as shown in the diagram.

193

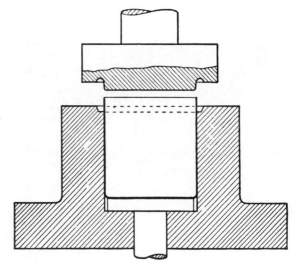

Fig. 7-8. A curling die.

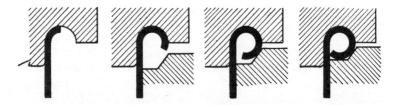

Fig. 7-9. Progressive operations in curling a bead around a cylindrical vessel.

In preparation for the wiring operation, a ring of wire is placed in position around the projecting edge of the workpiece. As illustrated in the progressive operations (Fig. 7-11), the punch, as it descends, coils the edge of the workpiece around the wire in successive steps.

Bulging Die

The basic principle of operation of the bulging die depends on the expensive action of a springy material placed under pressure. A rubber disk is subjected to pressure by a descending plunger (Fig. 7-12).

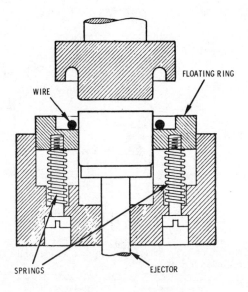

FLOATING RING

WIRE

SPRINGS

EJECTOR

Fig. 7-10. Diagram showing basic parts of a wiring die.

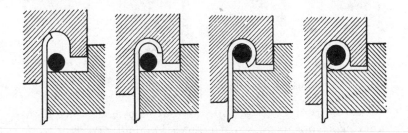

Fig. 7-11. Progressive operations of a wiring die.

In typical operations, the workpiece consisting of a shell drawn upright (see Fig. 7-12) is placed directly above the mushroom plunger, as shown in Fig. 7-13; and as the plunger descends, the rubber disk is expanded outward to expand or "bulge" the shell into the curved chamber formed by the punch and bulging die. A cylindrical projection is formed as the work enters and traverses the annular space.

195

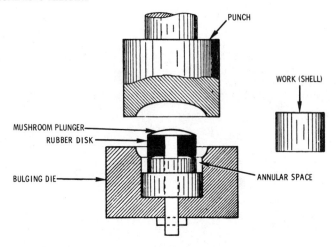

Fig. 7-12. Diagram of basic parts of a bulging die (left) and a section of the workpiece (right).

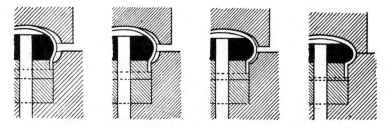

Fig. 7-13. Progressive operations of a bulging die.

After the progressive die operations have been completed, the workpiece appears as shown in Fig. 7-14. Similar types of work, such as covers or lids for some kinds of cans or containers, can be formed on the bulging die.

COMBINATION PUNCHING AND SHAPING DIES

Various combinations of punches and dies can be used to perform both the cutting and shaping operations. This is illustrated in the double-action die shown in Fig. 7-15.

196

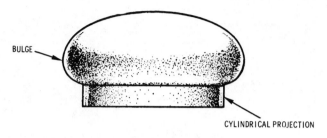

Fig. 7-14. Appearance of a typical workpiece that has been formed on a bulging die.

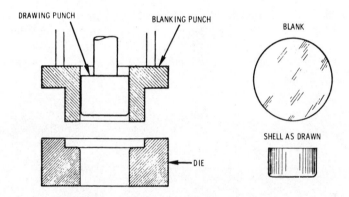

Fig. 7-15. A double-action die (left). Also shows the workpiece after blanking and drawing (right).

Double-Action Dies

Two concentric punches are used in this type of double-action die. An outer punch is used to perform the operations of cutting and holding the blank. The inner punch is a forming or drawing punch.

In actual operation (see Fig. 7-15), the cutting or blanking punch first descends to cut a blank. When the blank has been cut, the punch forces the blank onto the top of the die and presses it against the annular ring with enough pressure to prevent wrinkling; then it is allowed to recede into the forming part of the die during the drawing process (Fig. 7-16). Then, the inner or drawing punch descends to perform the drawing operation and finish the shell.

197

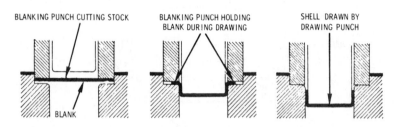

BLANKING PUNCH CUTTING STOCK

BLANKING PUNCH HOLDING
BLANK DURING DRAWING

SHELL DRAWN BY
DRAWING PUNCH

BLANK

Fig. 7-16. Sequence of progressive operations of the double-action blanking and drawing die.

The drawing dies usually employ a punch to shape or "draw" the work, and in some instances a second drawing operation, or "redrawing," is performed on work from the drawing die. Some typical drawn parts required by the electronics industry are shown in Fig. 7-17.

Courtesy Paul and Beekman, Inc.

Fig. 7-17. Some typical drawn parts for the electronics industry.

Plain Drawing Die

The plain drawing die is intended only for shallow drawing, as there is no pressure holder on the blank to prevent its wrinkling. A drawing punch having a diameter equal to the diameter of the hole minus twice the thickness of the blank or stock is illustrated in Fig. 7-18. The die hole is tapered slightly.

The blank fits in the recess at the top of the die. After the drawing operation is completed, the bottom edge of the die strips the shell from the punch on the upward stroke.

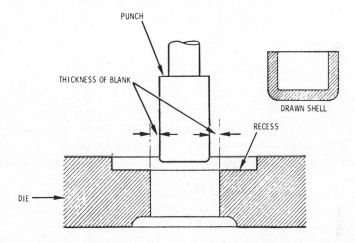

Fig. 7-18. A plain drawing die (left) and completed section of work (right).

Redrawing Die

The redrawing die performs a second operation on a workpiece that comes from the drawing die. A second drawing operation is performed, which reduces the diameter of the metal and increases its length. In other words, a cup or shell already formed from the stock is redrawn.

The redrawing die is similar to the drawing die in that the recess at the top of the die fits the shell that is to be redrawn, instead of being made to fit a blank that is to be drawn. A redrawing die with the shell in position for the redrawing operation is shown in Fig. 7-19.

199

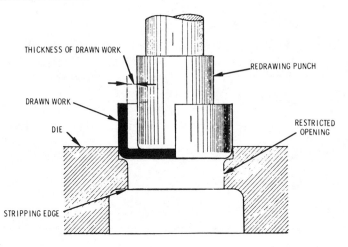

THICKNESS OF DRAWN WORK

REDRAWING PUNCH

DRAWN WORK

RESTRICTED OPENING

DIE

STRIPPING EDGE

Fig. 7-19. A redrawing punch.

Successive steps in the redrawing operation are shown in Fig. 7-20. During the downward stroke of the punch, the drawn shell is forced through the restricted opening. This squeezes the shell to a smaller diameter.

As the perimeter is reduced, the excess metal in the shell is squeezed lengthwise; therefore, the shell is lengthened by the redrawing operation. After the redrawing operation is completed, the workpiece is stripped or removed from the punch by the stripping edge.

Redrawing dies are sometimes called reducing dies because the redrawing operation reduces the diameter of the work. If considerable elongation is desired, the work is passed through a number of redrawing dies.

Gang and Follow Dies

A *gang die* is one in which two or more punches are combined in a single punching head. As many holes as there are mounted punches can be punched simultaneously in a single operation, that is, on the downward stroke of the assembly. A diagram of the basic parts of a gang die are shown in Fig. 7-21. The electronic chassis shown in Fig. 7-22 is an example of work done with a gang die.

200

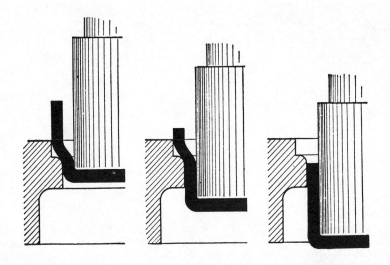

Fig. 7-20. Successive steps in the operation of the redrawing punch.

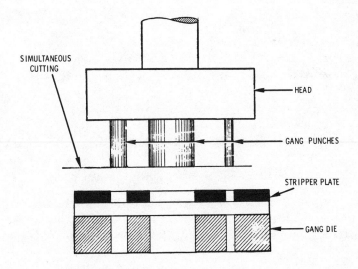

Fig. 7-21. A gang die. Note that all punches operate at the same time.

201

Fig. 7-22. An electronic chassis with a variety of punched holes.

A *follow die* is a modified gang die in which several successive operations are performed progressively. The difference between the gang die and the follow die is illustrated in Fig. 7-23.

Compound Die

The basic construction of a compound die is illustrated in Fig. 7-24. This type of die is constructed with a die in the upper punch

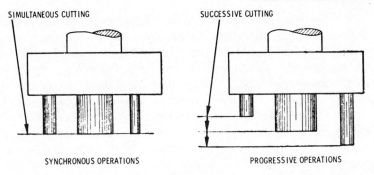

SIMULTANEOUS CUTTING

SUCCESSIVE CUTTING

SYNCHRONOUS OPERATIONS

PROGRESSIVE OPERATIONS

Fig. 7-23. Note the difference between a gang die (left) and a follow die (right). Note the progressive cutting operations of the follow die.

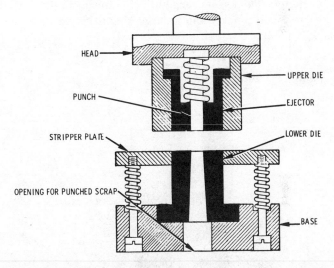

HEAD

UPPER DIE

PUNCH

EJECTOR

STRIPPER PLATE

LOWER DIE

OPENING FOR PUNCHED SCRAP

BASE

Fig. 7-24. Diagram showing basic construction of a compound die.

and a punch in the lower die; that is, it contains two punches and two dies. By the combined action of the two punches and the two dies, blanks are formed to the desired shape for which the die was designed.

Compound dies are different from plain and follow dies in that the basic punch and die elements are combined so that both the upper and lower members contain the equivalent of a punch and die, as well as stripper plates or ejectors.

203

In actual die operation (Fig. 7-25), the upper die descends and depresses the stripper plate. As the downward movement is continued, the blank is cut from the stock by the telescoping action of the upper and lower dies, the lower die acting as a punch. Both dies function as punches as well as dies; that is, the upper die acts as an external punch, and the lower die acts as an internal punch.

At the same time that the blank is cut, the punch pierces a central hole in the upper element. As the ram ascends, the blank is forced out of the upper die by the ejector, and the stripper plate pushes the work off the lower die. The metal punched from the hole falls through the opening in the lower die.

Delicate parts can be made with a compound die. Gear punchings for meters, clocks, etc., are typical of the accurate work that can be produced quickly and in large quantities with compound dies.

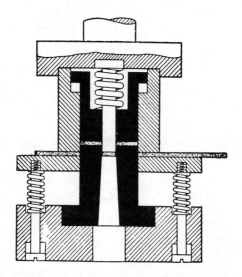

Fig. 7-25. Diagram of a compound die, showing the position at the end of the stroke.

Miscellaneous Dies

Numerous dies are made for various purposes. In addition to the dies already described, numerous other dies should be mentioned.

A *burnishing die* is constructed slightly smaller at the bottom than at the top. As the bottom of the die is a little smaller than the top, the work receives a high finish as it is forced downward through the die.

A *combination die* can be used to perform a combination of operations in a single stroke. The blank can be cut; and as the stroke progresses, the edge can be turned down and the piece drawn into the required shape.

A *trimming die* is used to remove excess metal. It is designed so that the excess metal is removed from the edges or ends of various drawn and formed workpieces.

The *fluid dies* are designed for forming fancy hollow ware from soft metals by means of hydraulic pressure. The shell that is to be formed is filled with liquid and enclosed within the die. Then, by means of a plunger, hydraulic pressure can be applied to force the shell outward against the contour of the enclosing walls of the die.

A *perforating die* is a multiple die. Many small-diameter punches can be grouped together to form the multiple die.

A *reducing die* is a type of redrawing die. It is designed to reduce only a portion of the shell instead of the entire length, as in regular redrawing dies.

Three independent movements can be made by the *triple-action die*. It can be used to perform such operations as cutting or blanking, drawing and stamping, or embossing.

Frequently, a part cannot be made in a single operation but requires several. Often these are made in such a way that the material is moved through a series of operations. These are known as *progressive dies*. Fig. 7-26 shows a 14-stage progressive die that forms the automobile part in the foreground. The workpiece in the center shows the various stages from the flat strip at the bottom to the finished piece at the top. The work is moved up one stage automatically with each stroke of the press.

DIEMAKING OPERATIONS

The choice of a suitable lubricant, selection of the proper material, and layout of the die are all important operations that must be considered in diemaking.

Fig. 7-26. A 14-stage progressive die.

Lubricants

Although some shops do not use a lubricant for working sheet metal, a lubricant is recommended for best results. A heavy animal oil can be used for cutting steel and brass. Lard oil can be used, but it is more expensive. A mixture of equal parts of oil and black lead can be used for drawing steel shells.

When working with German silver or copper stock, a thin coat of sperm oil can be used. Lard oil can also be used on these materials.

The lubricant can be spread over the stock by coating a single sheet heavily with the lubricant and passing it through a pair of rolls; then several sheets can be lubricated lightly by passing them through the rolls before the supply of lubricant on the rolls is exhausted. This is also a good method for lubricating work that is to be drawn because the thin coating of oil will disappear during the drawing operation, and cleaning the shells is unnecessary after

206

the operation has been completed. A brush or pad is objectionable for applying lubricant because the coating of oil is usually too thick and requires cleaning after the operation has been finished.

Clean soap and water is recommended as a lubricant for drawing brass and copper. For aluminum stock, lard oil, tallow, or vaseline is recommended.

Materials for Making Dies

In most instances, a good grade of tool steel is used for making punches and dies. The steel should be free of harmful impurities. Sometimes the body of the die can be made of cast iron with an inserted steel bushing to reduce the cost of material. An advantage of this type of construction is that an insert can be replaced when it becomes worn. Soft steel that has been case hardened does not change its form as readily as tool steel, and any minute changes in form can be corrected readily, as the interior is soft.

Internal strains or stresses are set up in steel during the manufacturing process. In diemaking operations, these stresses must be relieved before the die is brought to its final size, or they will cause distortion. The presence of stresses cannot be determined in the steel beforehand, but the diemaker can relieve the stresses in the steel by annealing after the die has been roughed out.

Laying Out Dies

Before laying out a die, the diemaker must determine the most economical way to punch the stock so that there will be a minimum waste of metal. This determines the layout for the die and the location of the gage or stop pin.

The simplest method of determining the layout is to cut several pieces to fit the outline of the blanks to be made. Then the pieces can be arranged in various ways to determine the most economical arrangement for punching the stock. If the problem has been solved correctly, the maximum number of blanks can be obtained from a given unit of stock.

Determining minimum waste—The cut-and-trial method can be used to determine the best layout for the diamond-shaped pattern (Fig. 7-27). Paper templates can be cut out and arranged

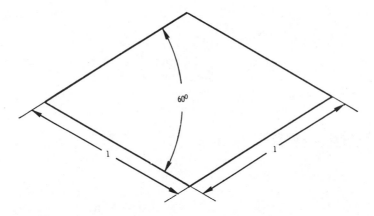

Fig. 7-28. Diamond-shaped pattern, used to illustrate a cut-and-trial method of determining the way to cut the stock for minimum waste of material.

on the sheet of stock, as shown in Fig. 7-28. Draw the enclosing rectangle *ABCD* to represent the amount of stock required to make four diamond-shaped blanks.

As can be noted in the illustration (see Fig. 7-28), the areas *1, 2,* etc., represent considerable waste metal or scrap. Then rearrange

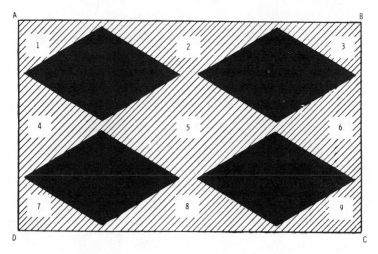

Fig. 7-28. Trial layout for the diamond-shaped pattern.

the templates (Fig. 7-29) and lay out another rectangle. Note that the rectangle *ABCD* is reduced to the rectangle *a b C′ D;* the saving in stock is represented by the rectangles *AB′ba* and *B′ B C C′*, leaving considerably less scrap metal when compared to the scrap metal left in the areas *1, 2,* etc., in Fig. 7-28. In this instance, the difference in waste can be determined by eye, but this is not always possible.

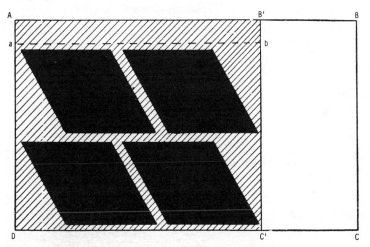

Fig. 7-29. A second trial layout for the diamond-shaped pattern. This is the correct layout for minimum waste of material.

Locating the gage pin—After positioning the workpiece for minimum waste of stock, the location of the gage pin should be determined. The gage pin should be located so that it will hold the workpiece in position, without shifting, when the stock is pushed against it.

A typical pattern (Fig. 7-30) can be used to illustrate a method of determining the location of the gage pin. In the illustration, the pattern is in the correct position for minimum waste of material.

The pin can be located with respect to the hole in the design or with respect to some other portion of its contour. As shown in Fig. 7-31, lay off the lines *ab* and *cd* to indicate the edges of the stock, and place the pattern in position with the point *B* as the center of the hole. Draw the axis *XX′* through the point *B*.

The position of the part the stock that forms the stop deter-

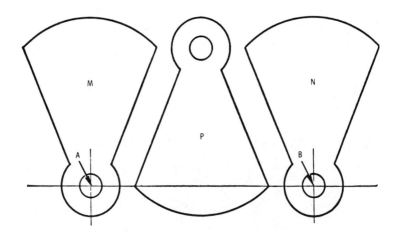

Fig. 7-30. A typical pattern for showing a method of determining the location of the gage pin.

mines the amount of stock that remains between the punched holes. The distance between one position of the stock and the gage pin can be determined from the patterns laid on the stock. To determine this distance, measure the distance on a line between corresponding points on the outlines of the two nearest patterns that occupy similar positions in reference to the edge of the stock. Thus, in Fig. 7-30, the patterns M and N occupy similar positions in relation to the edge of the stock, as opposed to the pattern P, which is reversed.

The gage pin is then located at a distance BA from the point B. Therefore, the distance BA (see Fig. 7-31) can be laid off on the line XX' equal to the distance BA in Fig. 7-30, which fixes the location of the gage pin in reference to the hole in the pattern.

A second layout is necessary to locate the gage pin with respect to some other part of the contour of the pattern (Fig. 7-32). If the gage pin is to be located within a space limited by the points r and t, draw the lines a a', b b', and c c' parallel to the edge of the stock and through the points r, s, and t. The points r', s', and t' on the face of the gage pin correspond to the points r, s, and t on the pattern, and they must be located at a distance AB (determined previously in Fig. 7-31) from the points r, s, and t, along the parallel lines $a\ a'$, $b\ b'$, and $c\ c'$.

The pattern P (see Fig. 7-30), which is placed 180° with respect to the patterns M and N, can be punched by turning the sheet of stock 180° and passing it through the press again. The best method of doing this in production work is to use a double blanking die, that is, a die that can punch the patterns M and P in a single operation.

Laying Out the Design on the Die

The surface of the stock from which the die is to be made should be finished smooth by either filing or grinding. Then the surface should be treated so that the scribed marks will be visible in transferring the design to the die.

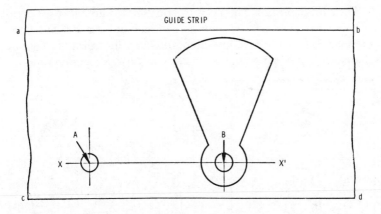

Fig. 7-31. Layout of pattern to locate the gage pin.

The surface should be free from grease. A solution of one part copper sulfate and ten parts water can be used to coat the surface. After a few minutes the solution will evaporate, leaving the surface covered with a thin film of copper, which serves to make finely scribed lines visible because of the difference in color between the steel and the copper.

211

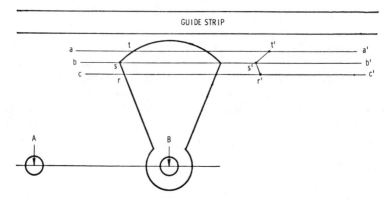

Fig. 7-32. A second layout for location of the gage pin with respect to some other point in the outline of the pattern.

Making the Die

The simple design (Fig. 7-33) can be used to illustrate the operations necessary to make the opening in a blanking die. The large circular ends A and B should be bored out with a drill of the same diameter. Then several small holes can be drilled to outline the

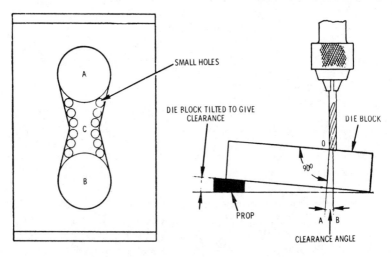

Fig. 7-33. Drilling operations for making the opening in a blanking die (left), and method of tilting the die block to obtain clearance (right).

intermediate part of the design. Alternate small holes should be drilled first; then the remaining holes should be drilled so that the entire metal section *C* can be removed from the design without difficulty.

A blanking die must have clearance (see Fig. 7-33). Clearance is the taper given the walls of the holes in a die—the large end of the hole is at the bottom of the die; this permits the punched blanks to fall out of the die. In Fig. 7-34, a die having no clearance is compared with a die having an exaggerated clearance.

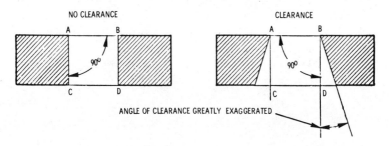

Fig. 7-34. Sections of die blocks having no clearance (left) and blocks having clearance (right).

Several methods can be used to obtain clearance. If there were no clearance, the punched blank would remain stuck in the die block (Fig. 7-35). The blank is free to fall out of the die block if the die has clearance.

In drilling to the outline of the die design, clearance can be obtained by inserting a prop or thin strip of metal beneath the block on the side farthest removed from the hole that is to be drilled (see Fig. 7-33). This tilts the die block so that the drilled hole will be inclined from the vertical at an angle equal to the clearance angle *AOB*.

The holes can be drilled at an angle of 90° to the face of the die block and then reamed from the bottom with a tapered reamer to provide the clearance angle. The opening at the bottom can also be made larger by filing the die opening to the scribed lines.

Usually, a clearance angle of 1° to 2° must be provided. The *diemakers' square* (Fig. 7-36) can be used to check for uniform clearance around the die opening. The diemakers' square differs

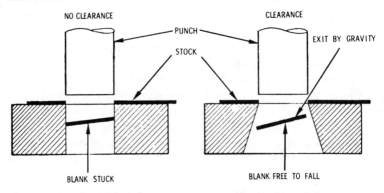

Fig. 7-35. Comparison of clearing actions of a die with no clearance (left) and a die with proper clearance (right).

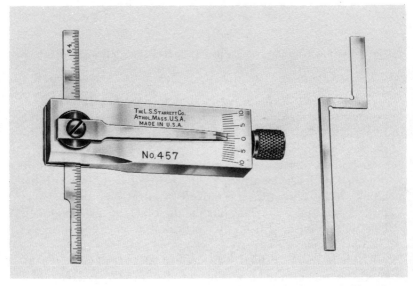

Fig. 7-36. A diemakers' square, used for measuring die clearances. The blade is locked in position by a small clamp screw, and can be set at any angle up to 8° on either side of 0°. The offset blade is used where it is impossible to sight with the straight blade.

from the real square in that the angle of the blade with the stock of the "square" is 92° (90° + 2°) to provide for the clearance angle. The blade is very narrow, for accessibility to small openings.

Surplus metal is removed from large dies by machining. It can be removed from the smaller dies with a sharp cold chisel after drilling. Clearance should not extend entirely upward to the cutting edge (upper face) of the die—it should extend only to within about ⅛ inch of the edge. This is because the die opening would be increased by sharpening, not because of the cutting action of the die.

Hardening and Tempering

The die is ready for hardening and tempering after the preceding operations have been completed. Exceptional skill and judgment are required to harden dies properly.

All holes, such as screw holes for the guide strips and the hole for the gage pin, should be plugged before beginning the hardening operation. Fire clay or asbestos can be used to plug these holes.

It is essential that a uniform heat be maintained during the hardening process. The die should be heated very slowly because the edges and portions having lesser quantities of metal will heat more rapidly than the rest of the die. If the die is heated unevenly, the die will contract unevenly during quenching, which tends to cause the die to crack. The die should be left in the quenching solution until it cools to room temperature. A strong salt brine can be a good quenching solution because it will harden the die satisfactorily at low heat. Salt water or brine will harden a die more satisfactorily at a lower temperature because the brine has a much greater conductivity than fresh water; consequently, it will absorb the heat much faster from the die. As the hardening of steel depends on the rapidity with which the heat is removed from the steel, it is reasonable to assume that a given degree of hardness can be obtained at a lower temperature when a brine solution is used as a quench.

After the die has been hardened, it should be brightened on its upper surface and tempered. The die can be tempered by placing it on a heated iron plate (¼ to ⅜ inch in thickness) with the bottom side of the die block downward on the plate. Uneven heating can be avoided by constantly moving the die. The color

215

for drawing depends on the kind of stock to be punched; usually the die should be quenched when the face of the die turns a deep straw color. Dies that are used for light-duty work can be harder than those used for heavy-duty work.

SUMMARY

The die consists of the die block and the punch and is used to cut or shape sheet metal or similar materials. The die block is the part of the die that contains a hole that has the same outline as the piece that is to be punched. The punch fits into the hole in the die block, which is the part to which power is applied for performing the cutting operation.

Dies can be classified with respect to the basic operation that they perform as cutting dies, shaping dies, and combined cutting and shaping dies. The cutting dies are the simplest form. The plain or blanking die is the type most frequently used. Depending on the shape of the punch, the blank can have any shape.

Various combinations of punches and dies can be used to perform both the cutting and shaping operations. Double-action dies, plain drawing dies, redrawing, gage and follow, and compound dies are just a few used in these operations. Two concentric punches are used in the double-action die. An outer punch is used to perform the operation of cutting and holding the blank. The inner punch is used as a forming or drawing punch. The plain drawing die is intended only for shallow drawing since there is little or no pressure on the blank.

Complicated sheet metal parts are often made by the use of progressive dies. Die clearances are checked with the diemakers' square. Heat treating a die requires exceptional skill and judgment.

REVIEW QUESTIONS

1. What is a die?
2. What are the three classifications of dies?
3. Name a few of the various types of dies.

4. What is a combination die and punch?
5. What is a stripper plate?
6. What are the basic parts of a die?
7. What is a double-action die?
8. Explain the operation of a compound die.
9. Why doesn't the clearance of a die extend to the cutting edge?
10. What is a progressive die?

CHAPTER 8

Grinding

The reader will want to refer to Volume I of the Machinists Library, where the subjects of abrasives, grinding wheels and the sharpening of drills, and lathe and planer tools are covered. It is assumed that the reader is knowledgeable concerning such material, so this chapter concentrates on the use of grinding wheels on various types of grinders.

Specific grinding machines are made for certain classes and sizes of work. Production grinders are of four main types: cylindrical center, cylindrical centerless, internal, and surface-grinding machines. The tool and cutter, thread, gear, and cam grinders are other types of grinding machines for special purposes.

CYLINDRICAL GRINDERS

Most external grinding of cylindrical parts is performed on cylindrical grinders. Cylindrical grinding designates a general category of grinding methods with the common characteristic of rotating the work around a fixed axis while grinding surface sections in relation to the axis of rotation. The part being ground is frequently cylindrical, hence the naming of the general category. Straight cylinders, cylinders having more than one diameter, and tapered parts can be ground on these grinders (Fig. 8-1). Even a surface of curvilinear profile may be ground as long as the condition of a common axis of rotation is satisfied. Cylindrical grinding is applied to manufacture of different sizes of work from miniature parts to rolling mill rolls weighing several tons. Pieces of work having irregular profiles, such as cams or eccentrics, are ground either on a cam grinder or a cylindrical grinder with a cam-grinding attachment.

Cylindrical work is usually held between centers. Live centers are used when the work is of such a nature that it would score either center, but two dead centers are used wherever possible. The centers are held in a headstock and tailstock, both mounted on a table that can be moved back and forth in front of the wheel.

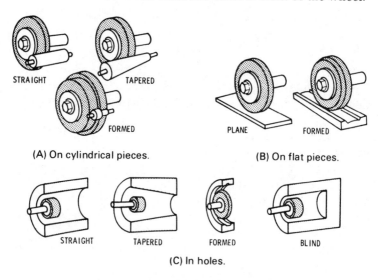

(A) On cylindrical pieces.

(B) On flat pieces.

(C) In holes.

Courtesy Cincinnati Milacron Co.

Fig. 8-1. Types of grinding commonly performed on external, internal, and flat work. Centerless grinding, either external or internal, is also performed on cylindrical work.

Cylindrical grinding machines are usually divided into three categories: (1) plain, (2) universal, and (3) limited-purpose. A plain cylindrical grinding machine is the basic type in this general category. It is used for grinding parts with cylindrical or slightly tapered form.

The universal cylindrical grinding machine, shown in Fig. 8-2, is a flexible machine. Fig. 8-3 shows a woodcut of the original universal grinder invented by J. R. Brown in 1876. The fundamental principles of the machines are the same. The most noticeable difference is the use of electric motors instead of the belt drives. In addition to the basic cylindrical forms, the universal grinder is useful for grinding steep tapers, face grinding, and even internal grinding. Tapers can be ground by swiveling the table, as illus-

Fig. 8-2. Valumaster Universal Grinder.

Fig. 8-3. Original universal grinder.

trated in Fig. 8-4, by swiveling the headstock, or by the use of special attachments such as a center grinding attachment. The wheel head can be swiveled and the wheel dressed to grind to a shoulder, or, in some cases, both the cylinder and a small shoulder, as illustrated in Fig. 8-5. Such variety is typical of work in the tool room, where most of these machines are installed.

Limited-purpose cylindrical grinders are basically production machines designed for special work and for high-volume production. Examples include crankshaft and camshaft grinders, roll grinders, etc.

If the work is long and slender, workrests may be necessary to support the work against the feed or pressure of the grinding wheels. They should be spaced evenly, from six to ten work

Fig. 8-4. Taper grinding on the universal grinding machine.

Fig. 8-5. Grinding to a shoulder.

diameters apart. This infeed and other basic data are listed in Table 8-1. The length of the surface to be ground is generally limited only by the movement of the machine table. In some larger roll grinders the movement is accomplished by the traverse of the wheel slide along a stationary machine table.

Table 8-1. Basic Process Data for Cylindrical Grinding

			TRAVERSE GRINDING			
Work Material	Material Condition	Work Surface Speed fpm	Infeed, Inch/Pass		Traverse for Each Work Revolution, In Fractions of The Wheel Width	
			Roughing	Finishing	Roughing	Finishing
Plain Carbon Steel	Annealed	100	0.002	0.0005	$\frac{1}{2}$	$\frac{1}{6}$
	Hardened	70	0.002	0.0003 to 0.0005	$\frac{1}{4}$	$\frac{1}{8}$
Alloy Steel	Annealed	100	0.002	0.0005	$\frac{1}{2}$	$\frac{1}{6}$
	Hardened	70	0.002	0.0002 to 0.0005	$\frac{1}{4}$	$\frac{1}{8}$
Tool Steel	Annealed	60	0.002	0.0005 max.	$\frac{1}{2}$	$\frac{1}{6}$
	Hardened	50	0.002	0.0001 to 0.0005	$\frac{1}{4}$	$\frac{1}{8}$
Copper Alloys	Annealed or Cold Drawn	100	0.002	0.0005 max.	$\frac{1}{3}$	$\frac{1}{6}$
Aluminum Alloys	Cold Drawn or Solution Treated	150	0.002	0.0005 max.	$\frac{1}{3}$	$\frac{1}{6}$

	PLUNGE GRINDING	
Work Material	Infeed Per Revolution of The Work, Inch	
	Roughing	Finishing
Steel, soft	0.0005	0.0002
Plain carbon steel, hardened	0.0002	0.000050
Alloy and tool steel, hardened	0.0001	0.000025

Cylindrical grinding between centers falls into two categories: traverse and plunge grinding. Traverse grinding is used where the work is longer than the maximum width of the grinding wheel that can be mounted on the machine. The worktable on which the headstock and tailstock centers are located reciprocates past the grinding wheel. As the work passes the wheel, a fixed amount of metal (infeed) is removed from the diameter. The grinding wheel is advanced another increment of distance, at the end of the pass, to remove the same amount of stock on the succeeding pass.

Fig. 8-6 shows a workpiece supported by two such workrests on a universal grinding machine. A simple method is to locate the first rest in the center of the work and use an equal number of rests on either side.

Courtesy Brown & Sharpe Manufacturing Co.

Fig. 8-6. Workrests being used on a universal grinding machine.

Plunge grinding is accomplished by using a grinding wheel that is at least as wide as the area to be ground. Grinding is accomplished by feeding the grinding wheel into the work, which is mounted between centers and does not reciprocate past the wheel. The infeed is referred to as the "infeed rate" because the infeed of the wheel is continuous rather than intermittent. This infeed rate, as shown in Table 8-1, is measured in ten-thousandths to millionths of an inch per revolution of the work.

Plunge grinding is more widely used in mass production grinding because it is more rapid and lends itself to automatic operation. These grinding wheels are usually ordered to specific thicknesses and are usually trued on the sides, as well as the face, to maintain the precise thickness. Production grinding machines are often built to mount more than one grinding wheel for plunge grinding several diameters on a single shaft simultaneously.

Cylindrical grinders are frequently fitted with grinding gages. These unique instruments consist of a dial indicator that reads the diameter of the work while the surface is being ground. The dial indicator can read in ten-thousandths of an inch, which allows the operator to grind right to size quickly and accurately. The gages can also be used with automatic machine controls. Fig. 8-7 shows a gage on the work being ground while Fig. 8-8 shows the gage retracted for replacing the workpiece.

CENTERLESS GRINDERS

Centerless grinding is the method of grinding cylindrical surfaces without rotating the work between fixed centers. This method of grinding is accomplished by supporting the work between three fundamental machine components—the grinding wheel, the regulating or feed wheel, and the work blade. The grinding and regulating wheels rotate in the same direction, clockwise, while the work rotates in a counterclockwise direction, establishing a "climb cut" operation.

Centerless grinding has developed into one of the most important operations in industry. The centerless grinder has proved itself in both high-production and job shops. The centerless grinding machine (Fig. 8-9) of the Cincinnati Milacron Co. can obtain roundness within 0.000050 inch, finishes below 0.000010 inch, and size tolerances within 0.0001 inch at high production rates.

The centerless grinder can be used on a wide range of materials such as cork, glass, porcelain, wood, rubber, plastics, and the new

Courtesy Federal Products Corp.

Fig. 8-7. Grinding gage measuring work during grinding.

high-alloy steels, as well as the more common types of ferrous and nonferrous materials. As its name implies, the centerless grinder requires no center points; there is no need to locate and drill center holes in the workpiece.

Basic principles—The principal parts of a centerless grinder are: (1) grinding wheel, (2) regulating wheel, and (3) workrest. Suitable guides are used in the "throughfeed" work rest to lead the workpiece to the wheels and to receive it from the wheels, as well as proper means for supporting the workpiece during the grinding cut. All these elements can be arranged and combined in various ways, but the fundamental principle involved is always the same.

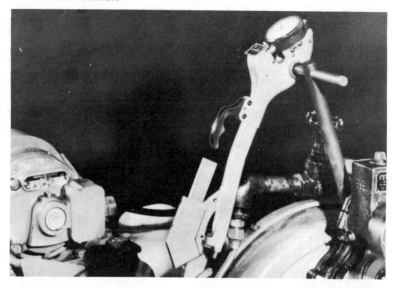

Fig. 8-8. Grinding gage retracted for replacing workpiece.

Fig. 8-9. Centerless grinding machine.

Action of grinding and regulating wheels—The action of the grinding wheel forces the workpiece against the workrest by means of *cutting pressure* and also against the regulating wheel by *cutting contact pressure*. The cutting pressure, aided by the force of gravity of the workpiece, keeps the workpiece in contact with the regulating wheel. The regulating wheel is usually made of material that is similar to the grinding wheel, and provides a continuously advancing frictional surface that insures constant and uniform rotation of the workpiece at the same velocity as the periphery of the regulating wheel.

It is rather simple to understand how a cylindrical surface can be ground on a center-type grinder. The diameter of the cylinder is governed by the distance between the line of centers and the face of the grinding wheel. However, there are no centers on the centerless grinder, and no apparent method of controlling the roundness of the workpiece exists. The ground diameter of the workpiece is determined by the distance between the two active surfaces of the wheels; however, a constant diameter does not necessarily indicate a perfect cylinder.

One of the simplest setups for centerless grinding is illustrated in Fig. 8-10. The center of the workpiece is in line with the centers of

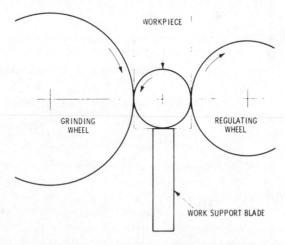

WORKPIECE

GRINDING
WHEEL

REGULATING
WHEEL

WORK SUPPORT BLADE

Fig. 8-10. Centerless grinding operation. Note the center point of the workpiece is placed "on center," and the work is supported by a blade with a flat top.

the grinding wheel and the regulating wheel. A work blade with a flat top is used to support the workpiece. The surfaces of the grinding wheel and the regulating wheel, together with the flat work blade, form three sides of a square.

The effect of an out-of-round piece of work placed "on center" on a work support blade with a flat top is shown in Fig. 8-11. A high

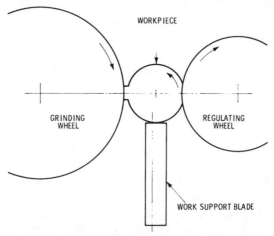

Fig. 8-11. The effect of centerless grinding an out-of-round workpiece placed "on center" on a work support blade with a flat top.

spot on the periphery of the workpiece produces a diametrically opposed concave spot when the high spot comes in contact with either the grinding wheel or the regulating wheel. A grinding setup of this type produces a piece of work that has a constant diameter, but the workpiece is not cylindrical. An exaggerated shape generated by this type of setup is illustrated in Fig. 8-12.

To correct the effect of an out-of-round workpiece, the center of the workpiece can be elevated above the center line of the wheels by raising the work support blade (Fig. 8-13). Then a low spot on the workpiece coming in contact with the regulating wheel causes a high spot to be generated at the point of contact with the grinding wheel, but the spots are not diametrically opposite. As the rotating piece is being ground, the low and high spots will not be opposite, and a gradual rounding effect is obtained. Maximum corrective action can be obtained by using a work support blade that is inclined at the top, as shown in the illustration.

230

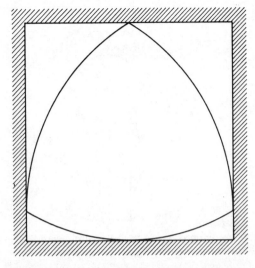

Fig. 8-12. An exaggerated shape resulting from centerless grinding an out-of-round workpiece placed "on center" on a work support blade with a flat top.

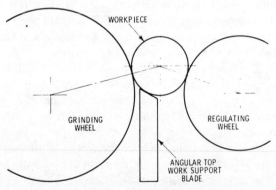

Fig. 8-13. Elevating the workpiece above the center line of the grinding and regulating wheels to correct the rounding action produced in grinding an out-of-round workpiece.

A diagrammatic sketch of the setup for corrective action in grinding an out-of-round workpiece is shown in Fig. 8-14. The two lines AA and BB are tangent at the points of contact of the workpiece with the grinding wheel and the regulating wheel, and another line CC indicates the plane of the inclined top of the work

231

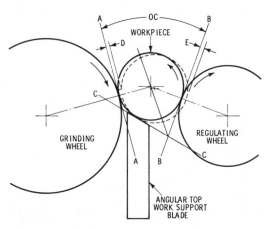

Fig. 8-14. Diagrammatic sketch of setup for corrective action in grinding an out-of-round workpiece.

support blade. If a low spot on the workpiece contacts either the work support blade or the regulating wheel (indicated by dotted lines), the approximate center point of the workpiece is lowered. As the center points of both wheels are fixed, the smallest possible diameter is generated only when the center point of the workpiece is located on their center line. Any increase in height of the center point of the work above the center line of the wheels causes the generation of a workpiece with a larger diameter, and lowering the center point of the work toward the center line of the wheels reduces the diameter of the ground piece. Thus the grinding wheel, instead of leaving a high spot on the periphery of the work corresponding to the depth of the concave spot at the point of contact with the regulating wheel, generates a proportionally smaller high spot at its point of contact with the workpiece. Theoretically, the low and high spots "dampen" themselves out, approaching the cylindrical shape in infinity, but the cylindrical form can be obtained in a short time in actual practice.

The position of the work in relationship to the two wheels should be such that the center of the work does not fall on the center line of the two wheels. The work is usually placed above center for fast rounding action. In some instances, especially for long, slender work, it is placed slightly below center to avoid a whipping action of the work.

232

Centerless grinders are used in shops where relatively large quantities of the same workpieces are ground. Some odd-shaped pieces, which tend to tilt when placed in the machine or to have projections too close to the ground surface, do not lend themselves to centerless grinding. There are other pieces, such as long, slender rods, which can be handled only on centerless machines.

Centerless grinding methods—Centerless grinding consists chiefly of four types: (1) throughfeed, (2) infeed, (3) endfeed, and (4) combined infeed and throughfeed.

Throughfeed—There is a fixed relation between the grinding wheel, regulating wheel, and the work support blade (Fig. 8-15). The regulating wheel can be adjusted so that the distance between the active surfaces of the wheel, together with the height of the work support blade, determines the diameter of the ground piece. The regulating wheel can be readjusted slightly to compensate for grinding wheel wear and restore the original relation between the wheels and support blade, which determines the size of the ground workpiece. The workrest, which supports the blade, is provided with adjustable guides both at the front and rear of the grinding wheel and regulating wheel. These guides direct the workpiece in a straight line to and from the wheels. Each of the guides can be adjusted by means of screws to the correct relation between the wheel surfaces and the diameter of the work.

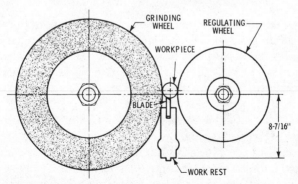

Fig. 8-15. The height of the wheel center points above the bottom of the workrest. The height of the center point of the workpiece relative to the center line of the wheels can be calculated by subtracting $8^7/_{16}$ inches from the distance between the center point of the workpiece and the finished surface of the lower slide.

233

In throughfeed grinding operations, the work is passed between the grinding and regulating wheels, and the grinding action takes place as the workpiece passes from one side of the wheels to the other side. Only *straight cylindrical surfaces without interfering shoulders* can be grounded by this method because all points on the workpiece pass all contact points between the wheels (Fig. 8-16).

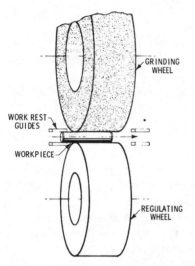

WORK REST GUIDES

WORKPIECE

GRINDING WHEEL

REGULATING WHEEL

Fig. 8-16. Top view of a through-feed centerless operation.

The regulating wheel imparts the axial movement of the work-piece past the grinding wheel. The regulating wheel can be swiv-eled about a horizontal axis from 2° below, to about 8° above, a line relative to the axis of the grinding wheel spindle. The speed and diameter of the regulating wheel also influence the feeding rate of the work.

Infeed—The infeed centerless grinding operation corresponds to plunge cut or form grinding on the center-type grinder. This method can be used to grind a workpiece *that has a shoulder, head, or other portion that is larger than the ground diameter* (Fig. 8-17). The infeed method can be used for simultaneous grinding of several diameters of a workpiece, as well as for finishing work-pieces that have an irregular profile. The length of the sections that can be ground in a single operation is not limited by the width of the grinding wheel. The combined infeed-throughfeed method

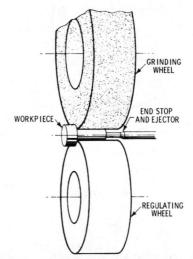

Fig. 8-17. Top view of an infeed centerless grinding operation.

can be used to grind work lengths that are longer than the width of the wheels.

There is usually (but not always) a fixed relation between the regulating wheel and the work support blade. The units that incorporate the regulating wheel and the work support blade are supported by two slides, which are clamped together (upper side clamps) and carry the workpiece to and from the grinding wheel. The infeed lever is turned 90° to perform this movement. As the lever is brought downward, the work and the regulating wheel are advanced to the grinding wheel, securing the desired size as the lever has completed its full swing. The gap between the wheels is increased as the movement of the infeed lever is reversed, and either a manually or automatically operated ejector kicks the work outward from between the wheels. Then another piece can be placed in position by the operator.

Form grinding is a good example of the type of infeed or plunge grinding on centerless machines. Fig. 8-18 illustrates this type of work. Out-of-balance work can sometimes be accomplished by the use of a hold-down shoe, as shown in Fig. 8-19.

Endfeed—This method of centerless grinding is used only on tapered workpieces. Either the grinding wheel or the regulating wheel, or both, must be dressed to the correct taper (Fig. 8-20). The grinding wheel, regulating wheel, and work support blade are

235

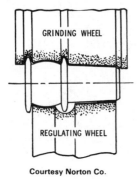

Fig. 8-18. Plunge grinding multiple-diameter work.

Courtesy Norton Co.

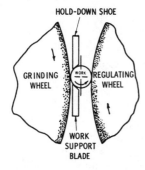

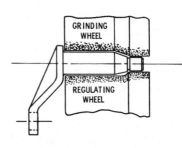

Courtesy Norton Co.

Fig. 8-19. Plunge grinding out-of-balance work using a hold-down shoe.

set in fixed relation to each other. The workpiece is fed inward from the front, either manually or mechanically, to a fixed end stop.

Combined infeed and throughfeed—The infeed and through-feed can often be combined (Fig. 8-21) for three chief types of applications as follows: (1) for grinding parts that can be ground more conveniently in a single pass but are too large for the conventional throughfeed method; (2) for grinding the smaller diameter of parts that have two diameters, where the portion that is to be ground exceeds the width of the grinding wheel; and (3) for warped parts in which the bow or warpage does not exceed the total amount of stock to be removed and the length of the part is less than the width of the grinding wheels.

Abrasive belt centerless grinding—Centerless grinding ma-

236

Fig. 8-20. Top view of an end feed centerless grinding operation.

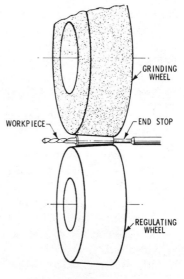

Fig. 8-21. Top view of combined infeed and throughfeed operation.

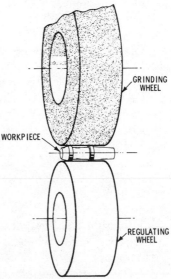

chines for cylindrical work have been designed using abrasive belt machines. Contact with the work is made at the point where the abrasive belt runs around one of the pulleys. The work is backed up by a regulating wheel. This wheel controls the rotation and

237

forward movement of the work just as in other centerless grinding machines. The finish and accuracy is comparable to the abrasive wheel centerless grinders for many types of straight work, and the belt machines are much less expensive.

Advantages of Centerless Grinding

There are several advantages offered by centerless grinding machines. Some of the chief advantages are:

1. As compared to center-type grinding, the loading time is small and enables the grinding process to be practically continuous.
2. The workpiece is rigidly supported directly beneath the grinding cut and for the full length of the cut. There is no deflection during the cut, and a heavier cut can be taken if desired.
3. Long, brittle workpieces and easily distorted workpieces can be ground because there is no axial thrust on the workpiece, as compared with center-type grinding.
4. The error of centering the workpiece is eliminated, as a true floating condition is present during the grinding process.
5. Possible error in setting up the job is reduced because stock removal is measured on the diameter rather than the radius; likewise, error due to wheel wear is reduced.
6. Maintenance is reduced because there are fewer wearing surfaces on the machine.
7. Larger quantities of small workpieces can be fed automatically by means of a magazine, a gravity chute, or hopper feeding attachments.
8. Extremely accurate control of size in production can be attained.
9. Very little skill is required by the machine operator.
10. Large grinding wheels can be used so that wheel wear is minimized.

INTERNAL GRINDING

Internal grinding machines are used for finishing holes to accurate diameters in such parts as bushings, gears, bearing races, cutters, and gages. Internal grinding may be done either on universal grinding machines (Fig. 8-22) or on machines especially designed for that purpose. Fig. 8-22 shows a setup for internal grinding on a universal grinding machine. The work is held in a chuck on the headstocks while the grinding wheel is mounted on an internal grinding spindle driven by a flat belt.

Courtesy Brown & Sharpe Manufacturing Co.

Fig. 8-22. Internal grinding on a universal grinder.

Rotating-work machine—In a machine of this type the work is held in a chuck on the spindle of the working head. The wheel spindle moves in and out of the hole during the grinding. Either straight holes or tapered holes may be ground as the headstock is made to swivel to various angles on either side of the center line.

Internal centerless grinding machine—Special machines have been developed for centerless internal grinding. The process is usually automatic, and the ground hole will be concentric with the outside diameter of the work. The machines can grind straight cylindrical or tapered holes. The holes may be blind, through, interrupted, or with a shoulder.

239

The principle is illustrated in Fig. 8-23. The process involves three rollers to hold the work and cause it to rotate on its outer surface. One roller is the regulating roller, which drives the work and controls its speed and direction. The second roller is mounted

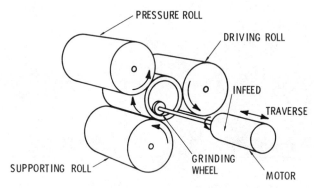

Fig. 8-23. Principle of centerless internal grinding.

below the work to support it and controls the distance from the work center. The third roller is a pressure roller and holds the work in contact with the other two. This roller moves in and out to allow for loading and unloading the machine. The grinding wheel is usually in a fixed position, and the work moves back and forth longitudinally. When two or more grinding operations must be performed on the same part, such as roughing and finishing, the work can be rechucked in the same location as often as required. The centerless grinding method simplifies the setup and permits automatic loading, so it is necessary only to keep the loading magazine full and to take away the ground parts. Fig. 8-24 shows two different procedures for grinding the inside of thin wall sections.

Cylinder grinding machine—stationary work—Sometimes called the planetary type of internal grinder, this machine was developed originally for the automobile industry for grinding cylinder block bores. The work is mounted on the table of the machine and travels to and from the wheel, which not only rotates on its axis, but also travels in a circular path of any desired diameter. This type of machine is now used for bore grinding in heavy or irregularly shaped work that cannot be readily rotated.

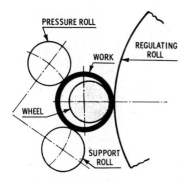

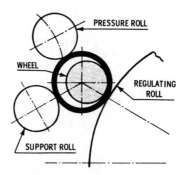

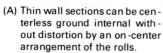

(A) Thin wall sections can be centerless ground internal without distortion by an on-center
arrangement of the rolls.

(B) High-center arrangement permits maintenance of high grinding accuracy, even with considerable variation in outside
diameter.

Courtesy Cincinnati Milacron Co.

Fig. 8-24. Internal centerless grinding of thin wall sections.

SURFACE GRINDERS

The operation of·producing and finishing flat surfaces with a
grinding machine is called surface grinding (Fig. 8-25). The term
"surface grinding" is generally accepted as meaning the grinding
of plane surfaces only. Surface grinders vary in size from small
toolroom machines using 7-inch × ½-inch wheels to very large
grinding machines with 48-inch to 60-inch segmental chucks (Fig.
8-26).

Surface grinders can be classified in two different ways—by the
type of table movement or by the position of the spindle. The table
movement is either reciprocating or rotary, and both types can be
obtained with horizontal or vertical spindles. The position of the
spindle is either vertical or horizontal, and either type can have a
reciprocating or rotary table.

Surface grinding requires that the spindle runs true so there is
constant pressure between the wheel and the work. If the spindle
bearings are loose, vibration of the spindle and wheel can happen,
which does not result in precision grinding.

The infeed or depth of cut is controlled by the amount of stock

(A) Horizontal spindle, reciprocating table.

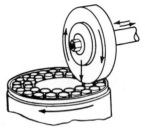

(B) Horizontal spindle, rotary table.

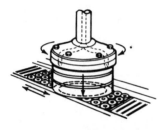

(C) Vertical spindle, recoprocating table

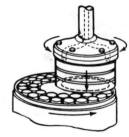

(D) Vertical spindle, rotary table.

(E) Horizontal spindle, single disk.

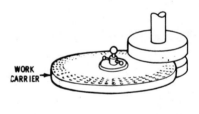

(F) Vertical spindle, double disk.

Courtesy Cincinnati Milacron Co

Fig. 8-25. Principle types of surface grinding include flat and rotary tables and horizontal and vertical wheels. Disk grinders may be single- or double-wheel and either vertical or horizontal.

to be removed, the desired finish, the operating temperatures, and the power capacity of the machine. A coolant can be used to reduce the operating temperature. An ample flow of coolant is more important than its pressure of application. Some vertical

242

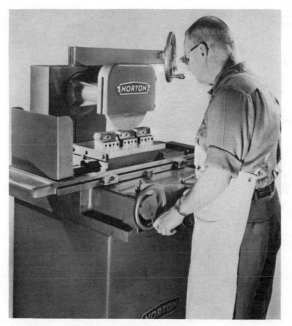

Courtesy Norton Co.

Fig 8-26. Hand-operated surface grinder.

spindle machines have the coolant fed through the hollow spindle to the inside of the grinding wheel. Wheel speeds are usually lower on vertical spindle machines because the area of contact between the work and the wheel is greater.

Planer-type surface grinders—The worktable reciprocates in this type of grinder (Fig. 8-27). There are three types of planer-type surface grinders: one uses the outside diameter, or periphery, of the grinding wheel mounted on a horizontal spindle; another uses a cup, cylinder, or segmental-type wheel mounted on a horizontal spindle; and a third uses a cup, cylinder, or segmental-type wheel mounted on a vertical spindle at right angles to the work.

Rotary-type surface grinders—There are two classes of rotary surface grinders: one uses the periphery of the grinding wheel mounted on a horizontal spindle (Fig. 8-28); and the other uses a cup, cylinder, or segmental wheel mounted on a vertical spindle.

243

Courtesy Norton Co.

Fig. 8-27. Horizontal-spindle, reciprocating-table surface grinder.

CUTTER AND TOOL GRINDING

A cutter deteriorates rapidly if it is used after it becomes dull. Its life will be prolonged if it is sharpened at the proper time because very little stock needs to be removed or ground off each time. The life and efficiency of a twist drill are greatly dependent on proper sharpening. Sharpening of lathe and planer tools likewise should receive careful attention and should be done by trained operators on suitable equipment with properly specified grinding wheels. It is especially important that multiple-point cutting tools, such as taps, milling cutters, hobs, reamers, etc., be sharpened properly.

Fig. 8-28. Rotary surface-grinding machine. Note that a taper-shank gear cutter is in position for grinding. To accommodate the shank, an adapter and support plate are mounted on the magnetic chuck, centering on a locator in the pilot hole.

Grinding Cemented Carbide Tools

The technique of grinding these tools has been developed to the point where they are considered no more costly to grind than high-speed steel tools. The expanded use of cemented carbides since World War II has emphasized the importance of correctly sharpening tools and cutters tipped with the hardest of cutting alloys.

245

Cutter Sharpening Machines

These machines are usually divided into two categories: (1) plain cutter grinders and (2) universal tool and cutter grinding machines. The plain cutter grinder is especially suitable for sharpening milling machine cutters and similar tools.

The universal tool and cutter grinding machines have sufficient range to meet practically all toolroom requirements. Basically, they are similar to small universal cylindrical grinders with four important differences:

1. No power feeds are provided; all table motions are manual.
2. The headstock can be swiveled about a horizontal axis as well as a vertical axis.
3. There is no motor on the headstock.
4. The wheelhead can be raised and lowered, as well as being swiveled 360° about a vertical axis.

The high degree of flexibility built into this machine makes it possible to grind almost any type of tool.

The grinding of cutting tools is primarily a problem of holding and moving the cutting tool in proper relationship to the grinding wheel. The setup of the grinder is a complicated job and requires a skilled workman. However, the actual grinding is rather easy after the machine is set up for a particular job. Fig. 8-29 shows the setup for grinding a cutter mounted on a mandrel. Sharpening such a cutter requires that the rake and clearance angles be correct. These are checked by use of a cutter clearance gage. Fig. 8-30 shows this gage being used to check the clearance angle of a milling cutter. Fig. 8-31 shows the same gage being used to check the clearance angle of a broach.

Carbide tools can be ground readily either with silicon carbide abrasive wheels (Fig. 8-32) or with diamond wheels, which may be resinoid, metal, or vitrified bonded. Aluminum oxide wheels are not suitable for grinding cemented carbide and should never be used on carbide tools, except for undercutting the steel shank.

Diamond grinding wheels are the accepted type of abrasive wheel for offhand finish grinding of carbide single-point tools and for all fixed feed or precision grinding operations on cemented carbides, including grinding of chip breakers and multitooth cut-

Fig. 8-29. Grinding a cutter on a universal tool and cutter grinder.

ters. They have the advantages of fast and cool cutting action and extremely low rate of wear, as compared to silicon carbide wheels. These advantages more than offset their higher price and make them more economical for these operations.

Diamond wheels, exclusively, are recommended for sharpening carbide-tipped face mills, reamers, and other multitooth cutters (Fig. 8-33). The wheels may be either vitrified or resinoid bonded.

BARREL FINISHING (ABRASIVE TUMBLING)

Barrel finishing is a precisely controlled method of processing quantities of parts, both metallic and nonmetallic, to remove sharp edges, burrs, machining or grinding lines, and heat-treat scale, and to improve surface finish (Fig. 8-34). The process may be used

247

Courtesy L.S. Starrett Co.

Fig. 8-30. Checking milling cutter clearance angle.

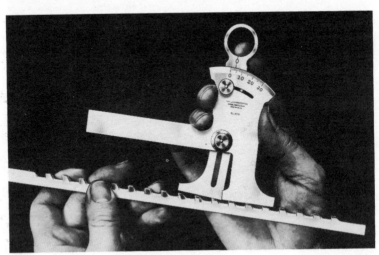

Courtesy L.S. Starrett Co.

Fig. 8-31. Checking broach clearance angle.

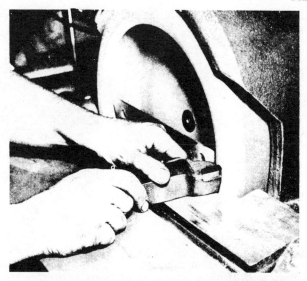

Courtesy Norton Co.

Fig. 8-32. Sharpening a single-point carbide planer tool on a silicon carbide grinding wheel.

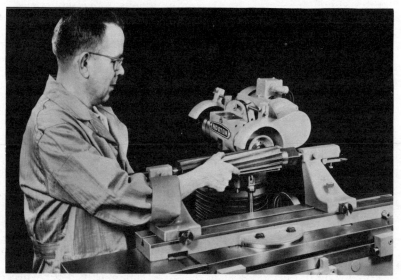

Courtesy Norton Co.

Fig. 8-33. Sharpening a reamer on a cutter and tool grinder.

Fig. 8-34. Before and after barrel finishing.

preliminary to buffing or plating operations. It may provide the desired finish for end use of the working parts for such machines as typewriters and business machines. The process is a modern quantity production method that results in savings in man-hours and cost per part manufactured.

Typical parts being finished commercially by the barrel-finishing method are stampings, die castings, small sand castings, machined parts, forgings, and sintered metals. Materials that can be barrel finished include steel, brass, cast iron, copper, bronze, aluminum, magnesium, titanium, silver, plastics, rubber, agate, and glass (Fig. 8-35).

Almost all barrel finishing in abrasive media is done wet. The essential components of the barrel-finishing process are:

1. A rotating barrel or vibration-type unit.
2. Abrasive media (a nonabrasive may be used for some purposes).
3. A chemical compound.
4. Water.

The basic action of barrel finishing is a sliding movement of the upper layer of the work load as the barrel rotates (Fig. 8-36). This action consists of a rotary movement of the mass of abrasive and parts to the point where the pull of gravity overcomes the ten-

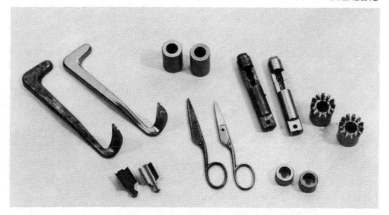

Courtesy Norton Co.

Fig. 8-35. Samples of typical parts before and after barrel finishing.

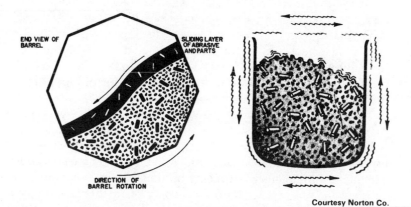

Courtesy Norton Co.

Fig. 8-36. Barrel-finishing action within a rotating barrel (left) and a vibra-
tory barrel (right).

dency of the mass to stay together. The upper layer of the load
then slides (neither falls nor is thrown) toward the lower portion of
the barrel. Nearly all (85 to 95 percent) of the abrading, deburring,
or polishing occurs during the sliding action. Little or no cutting
action takes place while the parts are moving up through the mass
from the bottom of the barrel to the point where the sliding starts.
A grinding, deburring, descaling, polishing effect, or honing to a
mirror finish can be secured by tumbling. The size and shape of

251

the barrel, the speed of rotation, the type of abrasive and lubricant, the proportion of parts and abrasive, the length of time, and the material and shape of the part being tumbled must be considered to secure the finish desired.

SUMMARY

Four main types of production grinders are: cylindrical center, cylindrical centerless, internal, and surface-grinding machines.

Cylindrical grinding is usually done between centers. The surface must be concentric around the axis but does not have to be parallel to the axis. Cylindrical grinding can be divided into two types—traverse and plunge grinding. The universal cylindrical grinding machine is quite flexible. Attachments can be used to provide additional operations. Gages can be used to actually measure the work while the part is being ground.

The basic parts of a centerless grinder are: grinding wheel, regulating wheel, and workrest. Various guides are used in the throughfeed workrest to lead the workpiece to the wheels and to receive it from the wheels, as well as a proper means of supporting the workpiece during the grinding cut.

The grinding wheel forces the workpiece against the workrest by means of the cutting pressure, and also against the regulating wheel by the cutting contact pressure. The regulating wheel is usually made of material that is similar to the grinding wheel and provides a continuously advancing frictional surface that insures constant and uniform rotation of the workpiece at the same velocity as the periphery of the regulating wheel.

Centerless grinding consists chiefly of four types: throughfeed, infeed, endfeed, and combined infeed and throughfeed. In throughfeed grinding operations, the work is passed between the grinding and regulating wheels, and the grinding action takes place as the workpiece passes from one side of the wheels to the other side.

The abrasive belt centerless grinder is an important development for many types of straight work. The finish and accuracy is as good as in the other type machines, while the belt machines are much less expensive.

Internal grinding is for finishing holes to accurate diameters. The operation can be done on universal grinding machines or on special machines. The centerless internal grinding machine has been a big cost saver in production. It is fully automatic and can produce small parts with the hole concentric with the outside diameter. And, best of all, it can produce the parts much cheaper than other methods.

Surface grinding is producing and finishing flat surfaces with a grinding machine. The spindle of a surface grinder can be in a horizontal or vertical position. The table movement can be either reciprocating or rotary.

Tool and cutter grinders are special grinders used to sharpen tools and cutters. The setup is complicated and requires a skilled toolmaker. The actual grinding is not difficult after the setup is made. The rake and clearance angles of tools and cutters need to be measured accurately.

Tumbling is a precisely controlled method of finishing quantities of parts. It is a vital part of the production process.

REVIEW QUESTIONS

1. Name the four types of grinding machines.
2. What are the three categories of cylindrical grinding machines?
3. Name the two categories of cylindrical grinding between centers.
4. What is centerless grinding?
5. What are the principle parts of a centerless grinder?
6. What are the four centerless grinding methods?
7. What are some of the advantages of centerless grinding?
8. What is the main advantage of abrasive belt centerless grinding?
9. Explain the principle of the centerless internal grinder.
10. What are the main advantages of the centerless internal grinder?
11. How are surface grinders classified?

12. Compare a universal tool and cutter grinder to a small universal cylindrical grinder.
13. How is the clearance angle on a cutter measured?
14. What are the advantages of barrel finishing?

CHAPTER 9

Laps and Lapping

In making tools, dies, jigs, fixtures, and gages, it is often necessary to provide still finer surface finishes and accuracy for certain surfaces after the grinding operation has been completed. On modern tools it is often necessary to use a lapping operation to obtain the finish necessary. Lapping is an abrading process done with a lap charged with an abrasive compound to: (1) produce true surfaces, (2) improve dimensional accuracy, (3) correct minor surface imperfections, and (4) provide a close fit between two surfaces. Gage blocks are lapped to plus or minus 0.000002 inch per inch of length, and they are also parallel within this dimension. Very little heat and pressure, which induce strains in the finished part, are involved in lapping, which is the reason it is more accurate than grinding and some other finishing operations.

Lapping is used on various types of surfaces—flat, cylindrical, spherical, or specially formed.

Lapping is accomplished by bringing the workpiece in contact with lap, usually made of a soft metal. Fine, loose abrasive mixed with a vehicle such as oil, grease, or water is used to charge the lap. Motion between the lap and workpiece is necessary to provide the abrading.

LAPS

A *lap* is a tool used for a superfine finish grinding or abrading operation. There are numerous types of laps designed to meet the requirements of various grinding operations.

Classification

Laps can be classified with respect to the degree of contact with the workpiece as *line* and *full* laps, and with respect to their shape as: (1) *cylindrical*, either ring or split; (2) *flat*; and (3) *adjustable*, which can be three-step, contracting, or expanding in type. Laps can also be classified with respect to their applications as either *internal* or *external*, and with respect to the material from which they are made as: (1) lead, (2) copper, (3) brass, (4) cast iron, and (5) steel.

Materials

Various materials are used for laps, depending on the service for which they are intended. A lap made of *lead* can be easily charged with abrasive, holds the abrasive firmly and does not scratch the workpiece. These laps are relatively easy to fit to the workpiece and hold their shape well for light cuts; they are also suitable for heavy-duty lapping operations.

Adjustment for wear can be provided on laps that are made of lead. The lap is molded to a tapered arbor, which provides for a slight adjustment (Fig. 9-1). The arbor should have a keyway to prevent the lap turning on the arbor. To adjust for wear, the arbor can be driven farther into the lap, thus expanding the lead. An internal adjustable split lap is illustrated in Fig. 9-2.

Of course, the lead will expand only until it fractures; therefore, the amount of adjustment for wear is quite limited. This expansion of the lead tends toward distortion, and the lap must be either

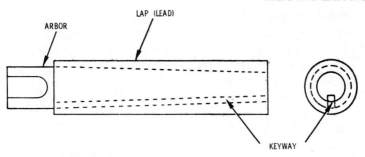

Fig. 9-1. An expanding internal lap made of lead (left). The lap is molded on the tapered arbor, which has a keyway (right) to prevent the lap turning on the arbor.

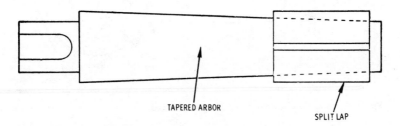

Fig. 9-2. An illustration of the internal adjustable split lap.

trued by turning or replaced after two or three adjustments. Laps made of lead have a disadvantage in that they tend to lose their form; however, they are inexpensive, easily molded, and quickly charged.

A lap made of close-grained *cast iron* is usually considered the best lap for fine accurate work. However, some machinists prefer a lap made of *copper* to either cast iron or lead. Laps that are made of copper can be charged more easily, and they cut more rapidly than cast-iron laps; but they do not produce as fine a finish.

Laps are also used to finish holes. Laps made of copper or brass are sometimes used for holes ¼ to ½ inch in size, and laps made of cast iron are used for the larger holes. Laps that are used to finish holes should be longer in length than the length of the hole.

External laps are usually made in the form of a *ring*. The ring forms the holder, and the inner shell is the lap proper. The inner shell can be made of lead, copper, etc., whichever material is best

257

suited for the purpose for which the lap is intended. The external lap is usually split, and adjustment is made by means of the screws in the holder (Fig. 9-3). The lap proper consists of a split ring inside a collar that is provided with a handle. A pointed screw engages the split ring to prevent its turning inside the collar. The remaining two screws are adjusted screws. The external lap should be at least as long as the diameter of the work.

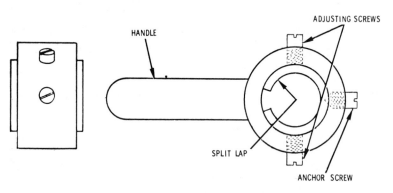

Fig. 9-3. Diagram of an external adjustable lap.

An adjustable external *step* lap for lapping plug gages is shown in Fig. 9-4. This type of lap is usually made of cast iron, and all holes are the same size. In the lapping operation, the first hole A is adjusted to size by means of the adjusting screw. Then the workpiece is passed from one hole to another during the lapping operation. There is less wear on the lap because of the several holes. The holes retain the same proportions, and they save time in frequent measuring checks when rough lapping the plug gage to within 0.001 inch of finished size.

Lapping Powders

The various "flours" used for lapping usually average about 0.002 inch in diameter. Lapping powders are made in a variety of sizes that indicate the grain or fineness of the abrasive particle used.

The ideal abrasive for lapping is one that will break down; that is, it will become more minutely divided during the lapping opera-

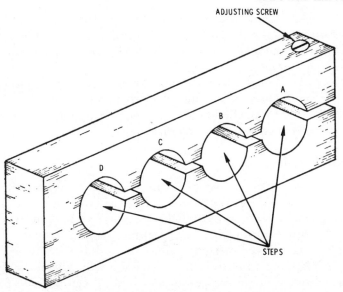

Fig. 9-4. An adjustable external step lap, used to lap plug gages.

tion. For fine work, the abrasives are mixed with oil and allowed to settle. The lighter particles remain in suspension and are poured off. The particles are graded as to the length of time that they float. The most-used lubricants for lapping operations are machine oil, hard oil, kerosine, gasoline, turpentine, and alcohol.

In recent years, diamond lapping powder has come into considerable use because of the necessity for grinding and lapping carbide tools to a fine finish. Diamond lapping powder is made by crushing small diamond chips in a steel mortar and then graining the dust for fineness. A coarse grade of powder is used for roughing tools; a fine grade is used for finishing tools.

LAPPING OPERATIONS

Lapping is a rubbing process for removing minute amounts of metal from surfaces that must be precision finished with respect to dimension and smoothness. A piece of work can be brought to a given dimension with greater accuracy and a finer finish than can

259

be obtained by any other means. *Note:* Laps were first used by lapidaries to finish the surfaces of mineral specimens, but laps have been in common use for some time in the machine shop to finish fine surfaces.

The more common lapping applications are those operations used to finish micrometer ends, plug and ring gages, holes in jig bushings, and for the finest die work. Lapping operations can be performed either by hand or by machine.

Hand Lapping

An important requirement of hand-lapping operations is that the motion should be in an ever-changing path to obtain uniform abrasion and to eliminate parallel grain marks on both the work and the lap. A matte (dull finish) surface can be obtained by lapping with an ever-changing path.

When lapping a flat surface, a No. 100 or No. 120 emery (or other abrasive of similar grade) and lard oil can be used for charging the roughing lap. After charging, the lap should be washed off and the surface kept moist with kerosene. Gasoline permits faster cutting speed, but the lap will not produce work that is true because of rapid and uneven evaporation of the gasoline.

Flat laps are usually made of cast iron, and they can be scored or grooved so that they will cut better for roughing (Fig. 9-5). A plain

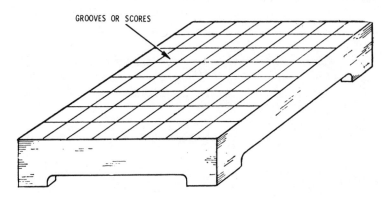

GROOVES OR SCORES

Fig. 9-5. A grooved flat lap made of cast iron.

260

lap with an unscored surface and charged with a fine abrasive can be used for a fine finish (Fig. 9-6).

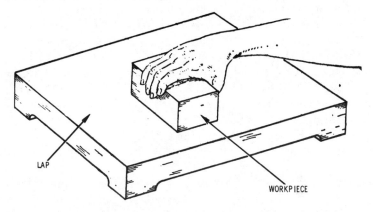

LAP

WORKPIECE

Fig. 9-6. A method of flat lapping.

Hand lapping is a slow and tedious operation. The operation is usually performed on flat surfaces by rubbing the parts to be lapped over the accurately finished flat surface of a master lap.

Machine Lapping

Three types of lapping medium are used on lapping machines in industry:

1. Metal laps and loose abrasive mixed with a lubricant.
2. Bonded abrasives for commercial production work.
3. Abrasive paper or cloth.

Vertical lapping machines with cast-iron laps and loose abrasives are used for both cylindrical and flat workpieces. The *universal* lapping machine can also be used for both types of work.

The vertical lapping machines are also used with bonded abrasive laps for both cylindrical and flat work. The horizontal lapping machines use either abrasive paper or cloth.

The workholders differ widely for the vertical lapping machines. The simplest form of workholders is a circular disk that is perforated with holes near its periphery.

These holes are made so that the work can be laid in them as the

261

workholder rests in position on the machine. Sufficient clearance is provided between the workpieces and the workholder for the workholder to move the work about between the laps.

When the workholder is arranged for cylindrical workpieces that do not have a hole through them, the openings are made to the approximate shape of the workpiece. The axis of the opening is not radial, but it is tangent to a circle and concentric with the center point of the workholder. A leg or spider-type workholder is used for cylindrical workpieces, such as piston pins, that have a hole through them.

The centerless lapping machine (Fig. 9-7) can be used to improve the accuracy and finish of commercially ground parts. Finishes can be obtained within 1 to 2 micro-inches; diameter tolerances within 0.00005 inch and 0.000025 inch are possible on the centerless lapping machine. Centerless lapping is particularly suitable for parts ground by the throughfeed centerless grinder. These parts can be lapped rapidly. When placed at the end of a line of centerless grinders, the centerless lapping machine can keep pace with the production rate of the grinders.

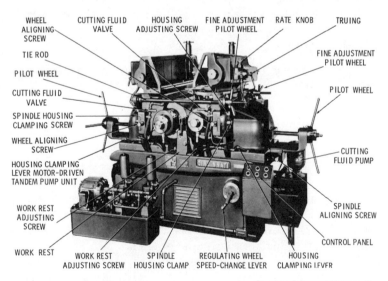

Courtesy Cincinnati Milacron Co.

Fig. 9-7. A centerless lapping machine.

Lapping a Cylinder

Engine cylinders that are only slightly out of round, or slightly tapered, can be trued up by lapping. Several different makes of electrical and mechanical lapping machines can be purchased on the market for lapping engine cylinders. The operation is continued in the cylinder until a measurement check with an inside micrometer or dial gage indicates that the cylinder is true for roundness and lack of taper. Then the lapping compound should be washed out of the cylinder bore and the new piston rings fitted. Sometimes it is necessary to lap a new piston lightly for it to fit in the cylinder properly.

Lapping a Tapered Hole

The often-used method of finishing tapered holes with a lap having the same taper as the hole is not recommended. The following procedure is recommended: (1) grind the hole to size, allowing for lapping; (2) without making any other changes, exchange the grinding wheel for a copper lap having the same shape as the grinding wheel but having a wider face; (3) lap in the same manner that was used in grinding the hole; and (4) avoid crowding the lap.

Rotary Disk Lap

A typical example of a job for a rotary disk lap is lapping the jaws of a snap gage (Fig. 9-8). The lap should be turned on the same arbor on which it is to be used, as it is practically impossible to return a lap to an arbor after it has been removed and then have it run true.

The disk lap (see Fig. 9-8) is made of cast iron, and the sides are relieved, leaving only a narrow edge or flange on each side to bear against the jaws. The lap is recessed deeply to allow for truing up each time that the lap is placed on the arbor. As the table is traversed back and forth, the lap passes over the entire surface of the gage jaw, grinding it in the same manner as with a cup-shaped grinding wheel. Care should be taken not to spring the snap gage as it is clamped in the vise.

The sides of the disk lap can be trued by clamping a keen-edged cutting tool in the vise, which is similar to the truing operation on a

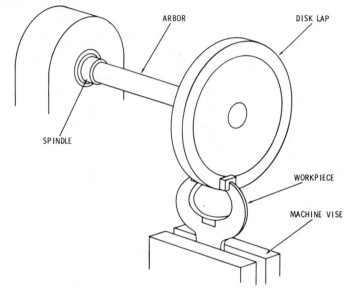

Fig. 9-8. Using a rotary disk lap to lap a snap gage.

lathe. This method is actually better than truing on the lathe because there is no change in the alignment of the lap with the work spindle. If the lap is true, a perfect contact between the lap and the gage is insured for the entire circumference of the lap.

A charging roller (Fig. 9-9) is used to charge the rotary disk lap. The abrasive is rolled in under moderate pressure. The charging roller should be made of a good grade of steel, and all loose abrasive should be washed off thoroughly after the lap has been charged.

SUMMARY

It is often necessary to improve the finish of certain surfaces after the grinding operation has been completed. On modern tools it is often necessary to use laps and lapping powders to obtain a fine finish and improve the accuracy.

A lap is a tool used for a grinding or an abrading operation. There are various types of laps designed to meet the requirements

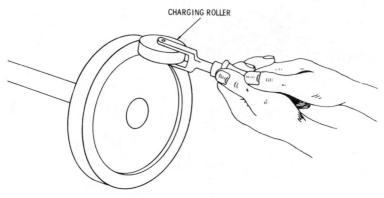

CHARGING ROLLER

Fig. 9-9. A method of charging a rotary disk lap.

of numerous grinding operations. Laps can be classified with respect to their application as either internal or external, and with respect to the material used in making the laps, such as lead, copper, brass, cast iron, and steel.

Lapping is a rubbing process for removing minute amounts of metal from surfaces that must be precision finished with respect to dimension and smoothness. Lapping operations can be performed either by hand or by machine.

REVIEW QUESTIONS

1. What is lapping?
2. What materials are used to make lapping tools?
3. What materials are used when lapping by hand?
4. What is the purpose of lapping, and why is it done?
5. What lubricants are used in lapping?

CHAPTER 10

Toolmaking Operations

A toolmaker is a master machinist who specializes in precision work, that is, dimensions of one-thousandth (0.001) of an inch or less. Mass production in this modern space age demands precision measuring. Recent developments have made the modern measuring tools more accurate and easier to read. Such things as improved finishes, wide-spaced verniers, and the expanded use of dial indicators, electronics, and optics have made measurements of a few millionths of an inch possible. Fig. 10-1 shows some of these precision measuring tools in use. However, many of these precise measurements are chiefly contact measurements.

Contact measurements require a sense of touch or "feel." A skilled toolmaker can detect a difference in dimension as small as 0.00025 inch. While this "feel" varies in individuals, it can be developed with practice. The measuring tool should be properly balanced in the hand and held lightly in such a way as to bring the fingers into play in adjusting or moving the tool. Fig. 10-2 shows

Fig. 10-1. Precision measuring tools in use.

the proper way an inside caliper is held. Note that a pair of "lock-joint transfer calipers" are being used.

INTRODUCTION

The basic or fundamental operations have been dealt with in the preceding books of this Machinists Library series. These operations must be mastered by any machinist who aspires to become a

268

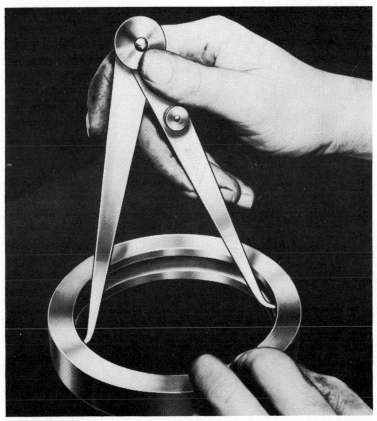

Courtesy L.S. Starrett Co.

Fig. 10-2. The proper way to hold an inside caliper.

master toolmaker—that is, a master of precision machining. Some of the requirements for becoming an expert toolmaker are:

1. Ability to read blueprints intelligently.
2. Knowledge of basic mathematics, including geometry and trigonometry.
3. Ability to read and to set precision measuring tools accurately and quickly.
4. Judgment as to the degree of precision required for the different tool parts.

269

5. Inventive ability—toolmaking constantly tests the ingenuity of the toolmaker.
6. Pride in keeping instruments and precision tools absolutely clean.

The art of toolmaking can best be acquired from experience. The apprentice machinist who desires to become a toolmaker practically begins a second apprenticeship in the toolroom, after completing an apprenticeship in the general shop.

At first the apprentice is permitted to perform the more simple operations of toolmaking. As skill and experience are gained, he can be trusted with more intricate work until he can master any work that comes into his department. A good toolmaker is always attempting to improve both his workmanship and his knowledge of tools and toolmaking; it is only after many years of work and study that he can consider himself an accomplished and expert toolmaker.

A good toolmaker is always a neat worker. He must handle drawings carefully, lay out his work painstakingly, and machine it accurately and carefully, keeping his machine clean and free from an accumulation of chips and grime. When setting up work and taking measurements, the toolmaker is careful to make sure that all bearing surfaces are clean. He also keeps the tools in his kit in perfect condition and adjustment because he knows that his workmanship is dependent on their accuracy.

ALLOWANCES AND TOLERANCES

The various methods of laying out and machining workpieces must be familiar to the toolmaker. The use of plugs or toolmakers' buttons must be understood thoroughly to avoid waste of time in accurately locating and boring holes. The toolmaker should possess a knowledge of the sine bar and its application to various forms of work, and he should be able to lay out and machine exacting angular types of work, such as dovetails, V-slides and V-ways, and plane angular surfaces.

The fundamental operations that are necessary in making tools have been presented at great length in previous volumes of the

Machinists Library series, treating the lathe, drill press, shaper, milling machine, etc. The purpose here is to deal with special requirements and problems related to precision that are encountered by the toolmaker, rather than the technique used to make the various tools.

Precision Measurements

Interchangeability and duplication of parts indicate to the toolmaker that the separate parts of a mechanism must be machined and finished so closely to the dimensions that they can be assembled into a complete unit without further fitting operations. The master toolmaker is always aware of this ideal. If further fitting is necessary, the parts cannot be said to be truly interchangeable. A minute error in dimensioning can require a certain amount of fitting as the parts are assembled into a complete machine unit.

A *precision measurement* must be accurate to within one-thousandth (0.001) part of an inch or fraction thereof. Such measurements are made with precision tools, which are constructed in such a manner that direct readings can be made of errors in dimension, even though the errors are minute.

Micrometers, verniers, etc., are examples of precision tools used for direct reading of minute measurements of lengths, diameters, depths, and heights. These tools are checked for accuracy to within ten-thousandths (0.001) of an inch. Micrometers are available that measure to ten-thousandths of an inch (0.0001 inch) or to two-thousandths of a millimeter (0.002 mm). Such instruments are usually checked with gage blocks that may be accurate within a few millionths of an inch, or plus or minus 0.00003 mm, or by the use of optical flats.

Tolerance Limits

In general practice, tolerances are certain allowances for error, minute or otherwise, that can be tolerated. While it is possible to produce machine parts with extremely accurate measurements, extreme precision can prove too costly for some types of commercial work.

271

As an example of a commercial practice to avoid waste of time, labor, and money, a set of rules has been formulated to define the degree of accuracy to be expected in instances where the specifications and drawings do not call for greater precision than that for which the rules provide. The set of rules is given here as follows:

1. Full information regarding limits of tolerance should be clearly shown by drawings submitted or be definitely covered by written specifications to which reference must be made by notations on the drawings.
2. Where the customer fails to supply proper data as to limits, this company's engineers will use their best judgment in deciding to what limits it is advisable to work. The company will not, in any event, assume responsibility for possible excessive cost brought about through working to closer limits than is necessary or for permitting greater latitude than may subsequently be found to be proper.
3. Where dimensions are stated in vulgar fractions with no limit of tolerance specified, it will be assumed that a considerable margin for variation from figured dimensions is available; unless otherwise ordered, the company's engineers will proceed according to the dictates of their best judgment as to what limits should be taken.
4. For all important dimensions, decimal figures should be used and limits clearly stated on detail drawings. If decimal figures are not used for such dimensions, a notation referring to the degree of accuracy required must be placed prominently on the drawing.
5. It is frequently necessary to reduce fractions representing fourths, eighths, sixteenths, thirty-seconds, and sixty-fourths to decimal equivalents. When a dimension of this character is expressed in a decimal equivalent and carried out to three, four, or five places, and if limits are not specified, it will be assumed that a limit of plus or minus 0.0015 is permissible, unless otherwise ordered.
6. Where dimensions are stated in decimal figures derived by other processes than those explained in paragraph 5, but with limits not specified, the following variations from dimensions stated can be expected:

Two-place decimals	0.005 plus or minus
Three-place decimals	0.0015 plus or minus
Four-place decimals	0.0005 plus or minus
Five-place decimals	0.0002 plus or minus

7. Where close dimensions, such as the location of holes from center to center in jigs, fixtures, machine parts, and other exact work of like character are required, detail drawings should be prominently marked "Accurate" and clear instructions be given.

8. The dimensions of internal cylindrical gages, external ring gages, snap gages, and similar work specified to be hardened, ground, and lapped will be obtained as accurately as the best mechanical practice applying to commercial work of the particular grade specified will permit.

9. As drilled holes vary in size from 0.002 inch to 0.015 inch (and in some instances even more) over the size of the drill used, those required to be made accurately to definitely specified sizes should be either reamed, ground, or lapped, and detail drawings thereof should bear notations accordingly.

10. American Standard Unified form of thread and pitches will be used unless other threads are specified.

Fits and Fitting

In machine construction, many parts bear such a close and important relation to one another that a certain amount of hand fitting is essential to make the surfaces contact properly. Fits can be classified as sliding, running, forced, and shrink fits, depending on their use.

If the surfaces that are in contact move on each other, the fit is a sliding or running fit. The fit is classified as a forced or shrink fit if the surfaces make contact with enough firmness to hold them together during ordinary usage.

If two surfaces are fitted to slide on each other without lost motion laterally, the fit is called a *sliding* fit. Examples of this class of fit are the cross and transverse slides of lathes, milling machines, drilling machines, boring machines, grinding machines, and planers.

In sliding fits, the moving and stationary parts of the machines are held in contact with each other by means of adjustable contact strips or *gibs*, sometimes called packing strips. The weight of the tables of grinding and planing machines can be great enough to keep the surfaces in close contact. The running bearings of the spindles, crankshafts, line shafts, etc., are examples of *running* fits.

In a *forced* fit, the parts are forced together under considerable pressure. Examples of this type of fit are the crankpins and axles in locomotive driving wheels and the cutterheads and spindles of numerous woodworking machines.

The amount of pressure required to force the two parts together is the limiting factor in forced fits. In "forcing" the axles into locomotive driving wheels, the specifications could limit the pressure to between 100 and 150 tons. However specified, it reduces to limits the size and requires the use of measuring tools. These measuring tools can be the direct-reading contact tools, such as the micrometer and vernier; or they can be the indirect-reading contact tools, such as the common spring caliper used in conjunction with thickness gages or "feelers."

In a *shrink* fit, the outer or enclosing member is expanded by heating before positioning the inner member. The contraction resulting from cooling brings the parts into firm contact.

Limits of Fits

If a pin or spindle is to be forced into a hole, or if a collar, hub, flange, or other machine part is to be shrunk onto a spindle or shaft, the diameter of the spindle is usually made larger to allow for fitting rather than decreasing the size of the hole. The amount to increase the diameter of the spindle or shaft depends on the length and diameter of the hole, the metal, or material, and the shape of the hub or collar to be fitted.

The American-British-Canadian conferences held in 1952 and 1953 resulted in a proposal for a system of limits and fits. A series of standard types and classes of fits on a unilateral hole basis has been developed, so that the fit produced by mating parts in a single class will produce a similar performance throughout the entire range of sizes. This system of fits and limits was revised in 1967. A series of standard tolerances is shown in Table 10-1.

Table 10-1. ANSI Standard Tolerances (ANSI B4.I-1967, Revised 1974)

Nominal Size, Inches	Grade									
	4	5	6	7	8	9	10	11	12	13
Over To	Tolerances in thousandths of an inch*									
0–0.12	0.12	0.15	0.25	0.4	0.6	1.0	1.6	2.5	4	6
0.12–0.24	0.15	0.20	0.3	0.5	0.7	1.2	1.8	3.0	5	7
0.24–0.40	0.15	0.25	0.4	0.6	0.9	1.4	2.2	3.5	6	9
0.40–0.71	0.2	0.3	0.4	0.7	1.0	1.6	2.8	*4.0	7	10
0.71–1.19	0.25	0.4	0.5	0.8	1.2	2.0	3.5	5.0	8	12
1.19–1.97	0.3	0.4	0.6	1.0	1.6	2.5	4.0	6	10	16
1.97–3.15	0.3	0.5	0.7	1.2	1.8	3.0	4.5	7	12	18
3.15–4.73	0.4	0.6	0.9	1.4	2.2	3.5	5	9	14	22
4.73–7.09	0.5	0.7	1.0	1.6	2.5	4.0	6	10	16	25
7.09–9.85	0.6	0.8	1.2	1.8	2.8	4.5	7	12	18	28
9.85–12.41	0.6	0.9	1.2	2.0	3.0	5.0	8	12	20	30
12.41–15.75	0.7	1.0	1.4	2.2	3.5	6	9	14	22	35
15.75–19.69	0.8	1.0	1.6	2.5	4	6	10	16	25	40
19.69–30.09	0.9	1.2	2.0	3	5	8	12	20	30	50
30.09–41.49	1.0	1.6	2.5	4	6	10	16	25	40	60
41.19–56.19	1.2	2.0	3	5	8	12	20	30	50	80
56.19–76.39	1.6	2.5	4	6	10	16	25	40	60	100
76.39–100.9	2.0	3	5	8	12	20	30	50	80	125
100.9–131.9	2.5	4	6	10	16	25	40	60	100	160
131.9–171.9	3	5	8	12	20	30	50	80	125	200
171.9–200	4	6	10	16	25	40	60	100	160	250

*All tolerances above heavy line are in accordance with American-British-Canadian (ABC) agreements.

Selection of the type of fit is based on the service required from the equipment being designed. The limits of size of the mating parts are established to be certain that the desired fit will be produced. Standard fits are designed by the following letter symbols as follows:

Running or sliding fit ....RC
Locational clearance fit LC
Locational transition fit ....LT
Locational interference fit ....LN
Force or shrink fit ....FN

Numbers representing the class of fit are also used in conjunction with the symbols representing the type of fit—for example, RC4 represents a Class 4, running fit. A general guide to machining

275

processes that can be used to produce work within the indicated tolerance grades is shown in Table 10-2.

Standard tolerance limits for American Standard running and sliding fits (RC), locational fits (LC, LT, and LN), and force and shrink fits (FN) are given in Tables 10-3 to 10-7.

Table 10-2. Machining Processes in Relation to Tolerance Grades

OPERATION	GRADE
Lapping and honing	4, 5
Cylinder grinding	5, 6, 7
Surface grinding	5, 6, 7, 8
Diamond turning	5, 6, 7
Diamond boring	5, 6, 7
Broaching	5, 6, 7, 8
Reaming	6, 7, 8, 9, 10
Turning	7, 8, 9, 10, 11, 12, 13
Boring	8, 9, 10, 11, 12, 13
Milling	10, 11, 12, 13
Planing and shaping	10, 11, 12, 13
Drilling	10, 11, 12, 13

LAYOUT

All dimensions should be given with special reference to the manner in which the tool is to be constructed. If the toolmaker must work from a certain surface in laying out the different parts of the tool and all measurements are made from that surface, the dimensions should be made to read from the surface (Fig. 10-3).

If irregularly spaced holes are to be located accurately, the dimensions should be given so that no calculations are required (see Fig. 10-3). It can be noted in the illustration that all dimensions are given with respect to the two surfaces, A and B.

For most operations the holes could be located approximately with ordinary tools. However, the distances between the holes must be given accurately for precision work, as when the toolmakers' buttons are to be used. If the distances between the holes are not given on the drawing, calculations that result in a loss of time are necessary.

For example, the three holes A, B, and C (see Fig. 10-3) require

Table 10-3. American National Standard Running and Sliding Fits (ANSI B4. I-1967, Revised 1974)

Values shown below are in thousandths of an inch

Nominal Size Range, Inches		Class RC 1			Class RC 2			Class RC 3			Class RC 4		
		Clearance*	Standard Tolerance Limits		Clearance*	Standard Tolerance Limits		Clearance*	Standard Tolerance Limits		Clearance*	Standard Tolerance Limits	
Over	To		Hole H5	Shaft g4		Hole H6	Shaft g5		Hole H7	Shaft f6		Hole H8	Shaft f7
0–	0.12	0.1 / 0.45	+0.2 / 0	-0.1 / -0.25	0.1 / 0.55	+0.25 / 0	-0.1 / -0.3	0.3 / 0.95	+0.4 / 0	-0.3 / -0.55	0.3 / 1.3	+0.6 / 0	-0.3 / -0.7
0.12–	0.24	0.15 / 0.5	+0.2 / 0	-0.15 / -0.3	0.15 / 0.65	+0.3 / 0	-0.15 / -0.35	0.4 / 1.12	+0.5 / 0	-0.4 / -0.7	0.4 / 1.6	+0.7 / 0	-0.4 / -0.9
0.24–	0.40	0.2 / 0.6	+0.25 / 0	-0.2 / -0.35	0.2 / 0.85	+0.4 / 0	-0.2 / -0.45	0.5 / 1.5	+0.6 / 0	-0.5 / -0.9	0.5 / 2.0	+0.9 / 0	-0.5 / -1.1
0.40–	0.71	0.25 / 0.75	+0.3 / 0	-0.25 / -0.45	0.25 / 0.95	+0.4 / 0	-0.25 / -0.55	0.6 / 1.7	+0.7 / 0	-0.6 / -1.0	0.6 / 2.3	+1.0 / 0	-0.6 / -1.3
0.71–	1.19	0.3 / 0.95	+0.4 / 0	-0.3 / -0.55	0.3 / 1.2	+0.5 / 0	-0.3 / -0.7	0.8 / 2.1	+0.8 / 0	-0.6 / -1.3	0.8 / 2.8	+1.2 / 0	-0.8 / -1.6
1.19–	1.97	0.4 / 1.1	+0.4 / 0	-0.4 / -0.7	0.4 / 1.4	+0.6 / 0	-0.4 / -0.8	1.0 / 2.6	+1.0 / 0	-1.0 / -1.6	1.0 / 3.6	+1.6 / 0	-1.0 / -2.0
1.97–	3.15	0.4 / 1.2	+0.5 / 0	-0.4 / -0.7	0.4 / 1.6	+0.7 / 0	-0.4 / -0.9	1.2 / 3.1	+1.2 / 0	-1.2 / -1.9	1.2 / 4.2	+1.8 / 0	-1.2 / -2.4
3.15–	4.73	0.5 / 1.5	+0.6 / 0	-0.5 / -0.9	0.5 / 2.0	+0.9 / 0	-0.5 / -1.1	1.4 / 3.7	+1.4 / 0	-1.4 / -2.3	1.4 / 5.0	+2.2 / 0	-1.4 / -2.8
4.73–	7.09	0.6 / 1.8	+0.7 / 0	-0.6 / -1.1	0.6 / 2.3	+1.0 / 0	-0.6 / -1.3	1.6 / 4.2	+1.6 / 0	-1.6 / -2.6	1.6 / 5.7	+2.5 / 0	-1.6 / -3.2
7.09–	9.85	0.6 / 2.0	+0.8 / 0	-0.6 / -1.2	0.6 / 2.6	+1.2 / 0	-0.6 / -1.4	2.0 / 5.0	+1.8 / 0	-2.0 / -3.2	2.0 / 6.6	+2.8 / 0	-2.0 / -3.8
9.85–	12.41	0.8 / 2.3	+0.9 / 0	-0.8 / -1.4	0.8 / 2.9	+1.2 / 0	-0.8 / -1.7	2.5 / 5.7	+2.0 / 0	-2.5 / -3.7	2.5 / 7.5	+3.0 / 0	-2.5 / -4.5
12.41–	15.75	1.0 / 2.7	+1.0 / 0	-1.0 / -1.7	1.0 / 3.4	+1.4 / 0	-1.0 / -2.0	3.0 / 6.6	+2.2 / 0	-3.0 / -4.4	3.0 / 8.7	+3.5 / 0	-3.0 / -5.2
15.75–	19.69	1.2 / 3.0	+1.0 / 0	-1.2 / -2.0	1.2 / 3.8	+1.6 / 0	-1.2 / -2.2	4.0 / 8.1	+2.5 / 0	-4.0 / -5.6	4.0 / 10.5	+4.0 / 0	-4.0 / -6.5

See footnotes at end of table.

Table 10-3. American National Standard Running and Sliding Fits (ANSI B4. I-1967, Revised 1974) (Continued)

Values shown below are in thousandths of an inch

Nominal Size Range, Inches Over — To	Class RC 5 Clearance*	Class RC 5 Hole H8	Class RC 5 Shaft e7	Class RC 6 Clearance*	Class RC 6 Hole H9	Class RC 6 Shaft e3	Class RC 7 Clearance*	Class RC 7 Hole H9	Class RC 7 Shaft d8	Class RC 8 Clearance*	Class RC 8 Hole H10	Class RC 8 Shaft c9	Class RC 9 Clearance*	Class RC 9 Hole H11	Class RC 9 Shaft
0- 0.12	0.6 / 1.6	+0.6 / 0	-0.6 / -1.0	0.6 / 2.2	+1.0 / 0	-0.6 / -1.2	1.0 / 2.6	+1.0 / 0	-1.0 / -1.6	2.5 / 5.1	+1.6 / 0	-2.5 / -3.5	4.0 / 8.1	+2.5 / 0	-4.0 / -5.6
0.12- 0.24	0.8 / 2.0	+0.7 / 0	-0.8 / -1.3	0.8 / 2.7	+1.2 / 0	-0.8 / -1.5	1.2 / 3.1	+1.2 / 0	-1.2 / -1.9	2.8 / 5.8	+1.8 / 0	-2.8 / -4.0	4.5 / 9.0	+3.0 / 0	-4.5 / -6.0
0.24- 0.40	1.0 / 2.5	+0.9 / 0	-1.0 / -1.6	1.0 / 3.3	+1.4 / 0	-1.0 / -1.9	1.6 / 3.9	+1.4 / 0	-1.6 / -2.5	3.0 / 6.6	+2.2 / 0	-3.0 / -4.4	5.0 / 10.7	+3.5 / 0	-5.0 / -7.2
0.40- 0.71	1.2 / 2.9	+1.0 / 0	-1.2 / -1.9	1.2 / 3.8	+1.6 / 0	-1.2 / -2.2	2.0 / 4.6	+1.6 / 0	-2.0 / -3.0	3.5 / 7.9	+2.8 / 0	-3.5 / -5.1	6.0 / 12.8	+4.0 / 0	-6.0 / -8.8
0.71- 1.19	1.6 / 3.6	+1.2 / 0	-1.6 / -2.4	1.6 / 4.8	+2.0 / 0	-1.6 / -2.8	2.5 / 5.7	+2.0 / 0	-2.5 / -3.7	4.5 / 10.0	+3.5 / 0	-4.5 / -6.5	7.0 / 15.5	+5.0 / 0	-7.0 / -10.5
1.19- 1.97	2.0 / 4.6	+1.6 / 0	-2.0 / -3.0	2.0 / 6.1	+2.5 / 0	-2.0 / -3.6	3.0 / 7.1	+2.5 / 0	-3.0 / -4.6	5.0 / 11.5	+4.0 / 0	-5.0 / -7.5	8.0 / 18.0	+6.0 / 0	-8.0 / -12.0
1.97- 3.15	2.5 / 5.5	+1.8 / 0	-2.5 / -3.7	2.5 / 7.3	+3.0 / 0	-2.5 / -4.3	4.0 / 8.8	+3.0 / 0	-4.0 / -5.8	6.0 / 13.5	+4.5 / 0	-6.0 / -9.0	9.0 / 20.5	+7.0 / 0	-9.0 / -13.5
3.15- 4.73	3.0 / 6.6	+2.2 / 0	-3.0 / -4.4	3.0 / 8.7	+3.5 / 0	-3.0 / -5.2	5.0 / 10.7	+3.5 / 0	-5.0 / -7.2	7.0 / 15.5	+5.0 / 0	-7.0 / -10.5	10.0 / 24.0	+9.0 / 0	-10.0 / -15.0
4.73- 7.09	3.5 / 7.6	+2.5 / 0	-3.5 / -5.1	3.5 / 10.0	+4.0 / 0	-3.5 / -6.0	6.0 / 12.5	+4.0 / 0	-6.0 / -8.5	8.0 / 18.0	+6.0 / 0	-8.0 / -12.0	12.0 / 28.0	+10.0 / 0	-12.0 / -18.0
7.09- 9.85	4.0 / 8.6	+2.8 / 0	-4.0 / -5.8	4.0 / 11.3	+4.5 / 0	-4.0 / -6.8	7.0 / 14.3	+4.5 / 0	-7.0 / -9.8	10.0 / 21.5	+7.0 / 0	-10.0 / -14.5	15.0 / 34.0	+12.0 / 0	-15.0 / -22.0
9.85-12.41	5.0 / 10.0	+3.0 / 0	-5.0 / -7.0	5.0 / 13.0	+5.0 / 0	-5.0 / -8.0	8.0 / 16.0	+5.0 / 0	-8.0 / -11.0	12.0 / 25.0	+8.0 / 0	-12.0 / -17.0	18.0 / 38.0	+12.0 / 0	-18.0 / -26.0
12.41-15.75	6.0 / 11.7	+3.5 / 0	-6.0 / -8.2	6.0 / 15.5	+6.0 / 0	-6.0 / -9.5	10.0 / 19.5	+6.0 / 0	-10.0 / -13.5	14.0 / 29.0	+9.0 / 0	-14.0 / -20.0	22.0 / 45.0	+14.0 / 0	-22.0 / -31.0
15.75-19.69	8.0 / 14.5	+4.0 / 0	-8.0 / -10.5	8.0 / 18.0	+6.0 / 0	-8.0 / -12.0	12.0 / 22.0	+6.0 / 0	-12.0 / -16.0	16.0 / 32.0	+10.0 / 0	-16.0 / -22.0	25.0 / 51.0	+16.0 / 0	-25.0 / -35.0

All data above heavy lines are in accord with ABC agreements. Symbols H5, g4, etc. are hole and shaft designations in ABC system. Limits for sizes above 19.69 inches are also given in the ANSI Standard.

*Pairs of values shown represent minimum and maximum amounts of clearance resulting from application of standard tolerance limits.

Courtesy Machinery's Handbook. The Industrial Press

Table 10-4. American National Standard Clearance Locational Fits (ANSI B4.1-1967, Revised 1974)

Values shown below are in thousandths of an inch.

Nominal Size Range, Inches, Over – To	Class LC 1 Clearance*	Class LC 1 Hole H6	Class LC 1 Shaft h5	Class LC 2 Clearance*	Class LC 2 Hole H7	Class LC 2 Shaft h6	Class LC 3 Clearance*	Class LC 3 Hole H8	Class LC 3 Shaft h7	Class LC 4 Clearance*	Class LC 4 Hole H10	Class LC 4 Shaft h9	Class LC 5 Clearance*	Class LC 5 Hole H7	Class LC 5 Shaft g6
0 – 0.12	0 / 0.45	+0.25 / 0	0 / -0.2	0 / 0.65	+0.4 / 0	0 / -0.25	0 / 1	+0.6 / 0	0 / -0.4	0 / 2.6	+1.6 / 0	0 / -1.0	0.1 / 0.75	+0.4 / 0	-0.1 / -0.35
0.12 – 0.24	0 / 0.5	+0.3 / 0	0 / -0.2	0 / 0.8	+0.5 / 0	0 / -0.3	0 / 1.2	+0.7 / 0	0 / -0.5	0 / 3.0	+1.8 / 0	0 / -1.2	0.15 / 0.95	+0.5 / 0	-0.15 / -0.45
0.24 – 0.40	0 / 0.65	+0.4 / 0	0 / -0.25	0 / 1.0	+0.6 / 0	0 / -0.4	0 / 1.5	+0.9 / 0	0 / -0.6	0 / 3.6	+2.2 / 0	0 / -1.4	0.2 / 1.2	+0.6 / 0	-0.2 / -0.6
0.40 – 0.71	0 / 0.65	+0.4 / 0	0 / -0.3	0 / 1.1	+0.7 / 0	0 / -0.4	0 / 1.7	+1.0 / 0	0 / -0.7	0 / 4.4	+2.8 / 0	0 / -1.6	0.25 / 1.35	+0.7 / 0	-0.25 / -0.65
0.71 – 1.19	0 / 0.7	+0.4 / 0	0 / -0.3	0 / 1.3	+0.8 / 0	0 / -0.5	0 / 2	+1.2 / 0	0 / -0.8	0 / 5.5	+3.5 / 0	0 / -2.0	0.3 / 1.6	+0.8 / 0	-0.3 / -0.8
1.19 – 1.97	0 / 0.9	+0.5 / 0	0 / -0.4	0 / 1.6	+1.0 / 0	0 / -0.6	0 / 2.6	+1.6 / 0	0 / -1	0 / 6.5	+4.0 / 0	0 / -2.5	0.4 / 2.0	+1.0 / 0	-0.4 / -1.0
1.97 – 3.15	0 / 1.0	+0.6 / 0	0 / -0.4	0 / 1.9	+1.2 / 0	0 / -0.7	0 / 3	+1.8 / 0	0 / -1.2	0 / 7.5	+4.5 / 0	0 / -3	0.4 / 2.3	+1.2 / 0	-0.4 / -1.1
3.15 – 4.73	0 / 1.2	+0.7 / 0	0 / -0.5	0 / 2.3	+1.4 / 0	0 / -0.9	0 / 3.6	+2.2 / 0	0 / -1.4	0 / 8.5	+5.0 / 0	0 / -3.5	0.5 / 2.8	+1.4 / 0	-0.5 / -1.4
4.73 – 7.09	0 / 1.5	+0.9 / 0	0 / -0.6	0 / 2.6	+1.6 / 0	0 / -1.0	0 / 4.1	+2.5 / 0	0 / -1.6	0 / 10.0	+6.0 / 0	0 / -4	0.6 / 3.2	+1.6 / 0	-0.6 / -1.6
7.09 – 9.85	0 / 1.7	+1.0 / 0	0 / -0.7	0 / 3.0	+1.8 / 0	0 / -1.2	0 / 4.0	+2.8 / 0	0 / -1.8	0 / 11.5	+7.0 / 0	0 / -4.5	0.6 / 3.6	+1.8 / 0	-0.6 / -1.8
9.85 – 12.41	0 / 2.0	+1.2 / 0	0 / -0.8	0 / 3.2	+2.0 / 0	0 / -1.2	0 / 5	+3.0 / 0	0 / -2.0	0 / 13.0	+8.0 / 0	0 / -5	0.7 / 3.9	+2.0 / 0	-0.7 / -1.9
12.41–15.75	0 / 2.1	+1.2 / 0	0 / -0.9	0 / 3.6	+2.2 / 0	0 / -1.4	0 / 5.7	+3.5 / 0	0 / -2.2	0 / 15.0	+9.0 / 0	0 / -6	0.7 / 4.3	+2.2 / 0	-0.7 / -2.1
15.75–19.69	0 / 2.4	+1.4 / 0	0 / -1.0	0 / 4.1	+2.5 / 0	0 / -1.6	0 / 6.5	+4 / 0	0 / -2.5	0 / 16.0	+10.0 / 0	0 / -6	0.8 / 4.9	+2.5 / 0	-0.8 / -2.4

*Clearance

See footnotes at end of table.

Table 10-4. American National Standard Clearance Locational Fits (ANSI B4. I-1967, Revised 1974) (Continued)

Values shown below are in thousandths of an inch

Nominal Size Range, Inches (Over – To)	Class LC 6 Clearance*	Class LC 6 Hole H9	Class LC 6 Shaft f8	Class LC 7 Clearance*	Class LC 7 Hole H10	Class LC 7 Shaft e9	Class LC 8 Clearance*	Class LC 8 Hole H10	Class LC 8 Shaft d9	Class LC 9 Clearance*	Class LC 9 Hole H11	Class LC 9 Shaft c10	Class LC 10 Clearance*	Class LC 10 Hole H12	Class LC 10 Shaft	Class LC 11 Clearance*	Class LC 11 Hole H13	Class LC 11 Shaft
0–0.12	0.3 / 1.9	+1.0 / 0	-0.3 / -0.9	0.6 / 3.2	+1.6 / 0	-0.6 / -1.6	1.0 / 2.0	+1.6 / 0	-1.0 / -2.0	2.5 / 6.6	+2.5 / 0	-2.5 / -4.1	4 / 12	+4 / 0	-4 / -8	5 / 17	+6 / 0	-5 / -11
0.12–0.24	0.4 / 2.3	+1.2 / 0	-0.4 / -1.1	0.8 / 3.8	+1.8 / 0	-0.8 / -2.0	1.2 / 4.2	+1.8 / 0	-1.2 / -2.4	2.8 / 7.6	+3.0 / 0	-2.8 / -4.6	4.5 / 14.5	+5 / 0	-4.5 / -9.5	6 / 20	+7 / 0	-6 / -13
0.24–0.40	0.5 / 2.8	+1.4 / 0	-0.5 / -1.4	1.0 / 4.6	+2.2 / 0	-1.0 / -2.4	1.2 / 5.2	+2.2 / 0	-1.6 / -3.0	3.0 / 8.7	+3.5 / 0	-3.0 / -5.2	5 / 17	+6 / 0	-5 / -11	7 / 25	+9 / 0	-7 / -16
0.40–0.71	0.6 / 3.2	+1.6 / 0	-0.6 / -1.6	1.2 / 5.6	+2.8 / 0	-1.2 / -2.8	2.0 / 6.4	+2.8 / 0	-2.0 / -3.6	3.5 / 10.3	+4.0 / 0	-3.5 / -6.3	6 / 20	+7 / 0	-6 / -13	8 / 28	+10 / 0	-8 / -18
0.71–1.19	0.8 / 4.0	+2.0 / 0	-0.8 / -2.0	1.6 / 7.1	+3.5 / 0	-1.6 / -3.6	2.5 / 8.0	+3.5 / 0	-2.5 / -4.5	4.5 / 13.0	+5.0 / 0	-4.5 / -8.0	7 / 23	+8 / 0	-7 / -15	10 / 34	+12 / 0	-10 / -22
1.19–1.97	1.0 / 5.1	+2.5 / 0	-1.0 / -2.6	2.0 / 8.5	+4.0 / 0	-2.0 / -4.5	3.6 / 9.5	+4.0 / 0	-3.0 / -5.5	5.0 / 15.0	+6 / 0	-5.0 / -9.0	8 / 28	+10 / 0	-8 / -18	12 / 44	+16 / 0	-12 / -28
1.97–3.15	1.2 / 6.0	+3.0 / 0	-1.0 / -3.0	2.5 / 10.0	+4.5 / 0	-2.5 / -5.5	4.0 / 11.5	+4.5 / 0	-4.0 / -7.0	6.0 / 17.5	+7 / 0	-6.0 / -10.5	10 / 34	+12 / 0	-10 / -22	14 / 50	+18 / 0	-14 / -32
3.15–4.73	1.4 / 7.1	+3.5 / 0	-1.4 / -3.6	3.0 / 11.5	+5.0 / 0	-3.0 / -6.5	5.0 / 13.5	+5.0 / 0	-5.0 / -8.5	7 / 21	+9 / 0	-7 / -12	11 / 39	+14 / 0	-11 / -25	16 / 60	+22 / 0	-16 / -38
4.73–7.09	1.6 / 8.1	+4.0 / 0	-1.6 / -4.1	3.5 / 13.5	+6.0 / 0	-3.5 / -7.5	6 / 16	+6 / 0	-6 / -10	8 / 24	+10 / 0	-8 / -14	12 / 44	+16 / 0	-12 / -28	18 / 68	+25 / 0	-18 / -43
7.09–9.85	2.0 / 9.3	+4.5 / 0	-2.0 / -4.8	4.0 / 15.5	+7.0 / 0	-4.0 / -8.5	7 / 18.5	+7 / 0	-7 / -11.5	10 / 29	+12 / 0	-10 / -17	16 / 52	+18 / 0	-16 / -34	22 / 78	+28 / 0	-22 / -50
9.85–12.41	2.2 / 10.2	+5.0 / 0	-2.2 / -5.2	4.5 / 17.5	+8.0 / 0	-4.5 / -9.5	7 / 20	+8 / 0	-7 / -12	12 / 32	+12 / 0	-12 / -20	20 / 60	+20 / 0	-20 / -40	28 / 88	+30 / 0	-28 / -58
12.41–15.75	2.5 / 12.0	+6.0 / 0	-2.5 / -6.0	5.0 / 20.0	+9.0 / 0	-5 / -11	8 / 23	+9 / 0	-8 / -14	14 / 37	+14 / 0	-14 / -23	22 / 66	+22 / 0	-22 / -44	30 / 100	+35 / 0	-30 / -65
15.75–19.69	2.8 / 12.8	+6.0 / 0	-2.8 / -6.8	5.0 / 21.0	+10.0 / 0	-5 / -11	9 / 25	+10 / 0	-9 / -15	16 / 42	+16 / 0	-16 / -26	25 / 75	+25 / 0	-25 / -50	35 / 115	+40 / 0	-35 / -75

All data above heavy lines are in accordance with American-British-Canadian (ABC) agreements. Symbols H6, H7, s6, etc. are hole and shaft designations in ABC system. Limits for sizes above 19.69 inches are not covered by ABC agreements but are given in the ANSI Standard.

*Pairs of values shown represent minimum and maximum amounts of interference resulting from application of standard tolerance limits.

Table 10-5. ANSI Standard Transition Locational Fits (ANSI B4. I-1967, Revised 1974)

Values shown below are in thousandths of an inch

Nominal Size Range, Inches Over To	Class LT 1 Fit*	Hole H7	Shaft js6	Class LT 2 Fit*	Hole H8	Shaft js7	Class LT 3 Fit*	Hole H7	Shaft k6	Class LT 4 Fit*	Hole H8	Shaft k7	Class LT 5 Fit*	Hole H7	Shaft n6	Class LT 6 Fit*	Hole H7	Shaft n7
0–0.12	−0.12 +0.52	+0.4 0	+0.12 −0.12	−0.2 +0.8	+0.6 0	+0.2 −0.2							−0.5 +0.15	+0.4 0	+0.5 +0.25	−0.65 +0.15	+0.4 0	+0.65 +0.25
0.12–0.24	−0.15 +0.65	+0.5 0	+0.15 −0.15	−0.25 +0.95	+0.7 0	+0.25 −0.25							−0.6 +0.2	+0.5 0	+0.6 +0.3	−0.8 +0.2	+0.5 0	+0.8 +0.3
0.24–0.40	−0.2 +0.8	+0.6 0	+0.2 −0.2	−0.3 +1.2	+0.9 0	+0.3 −0.3	−0.5 +0.5	+0.6 0	+0.5 +0.1	−0.7 +0.8	+0.9 0	+0.7 +0.1	−0.8 +0.2	+0.6 0	+0.8 +0.4	−1.0 +0.2	+0.6 0	+1.0 +0.4
0.40–0.71	−0.2 +0.9	+0.7 0	+0.2 −0.2	−0.35 +1.35	+1.0 0	+0.35 −0.35	−0.5 +0.6	+0.7 0	+0.5 +0.1	−0.8 +0.9	+1.0 0	+0.8 +0.1	−0.9 +0.2	+0.7 0	+0.9 +0.5	−1.2 +0.2	+0.7 0	+1.2 +0.5
0.71–1.19	−0.25 +1.05	+0.8 0	+0.25 −0.25	−0.4 +1.6	+1.2 0	+0.4 −0.4	−0.6 +0.7	+0.8 0	+0.6 +0.1	−0.9 +1.1	+1.2 0	+0.9 +0.1	−1.1 +0.2	+0.8 0	+1.1 +0.6	−1.4 +0.2	+0.8 0	+1.4 +0.6
1.19–1.97	−0.3 +1.3	+1.0 0	+0.3 −0.3	−0.5 +2.1	+1.6 0	+0.5 −0.5	−0.7 +0.9	+1.0 0	+0.7 +0.1	−1.1 +1.5	+1.6 0	+1.1 +0.1	−1.3 +0.3	+1.0 0	+1.3 +0.7	−1.7 +0.3	+1.0 0	+1.7 +0.7
1.97–3.15	−0.3 +1.5	+1.2 0	+0.3 −0.3	−0.6 +2.4	+1.8 0	+0.6 −0.6	−0.8 +1.1	+1.2 0	+0.8 +0.1	−1.3 +1.7	+1.8 0	+1.3 +0.1	−1.5 +0.4	+1.2 0	+1.5 +0.8	−2.0 +0.4	+1.2 0	+2.0 +0.8
3.15–4.73	−0.4 +1.8	+1.4 0	+0.4 −0.4	−0.7 +2.9	+2.2 0	+0.7 −0.7	−1.0 +1.3	+1.4 0	+1.0 +0.1	−1.5 +2.1	+2.2 0	+1.5 +0.1	−1.9 +0.4	+1.4 0	+1.9 +1.0	−2.4 +0.4	+1.4 0	+2.4 +1.0
4.73–7.09	−0.5 +2.1	+1.6 0	+0.5 −0.5	−0.8 +3.3	+2.5 0	+0.8 −0.8	−1.1 +1.5	+1.6 0	+1.1 +0.1	−1.7 +2.4	+2.5 0	+1.7 +0.1	−2.2 +0.4	+1.6 0	+2.2 +1.2	−2.8 +0.4	+1.6 0	+2.8 +1.2
7.09–9.85	−0.6 +2.4	+1.8 0	+0.6 −0.6	−0.9 +3.7	+2.8 0	+0.9 −0.9	−1.4 +1.6	+1.8 0	+1.4 +0.2	−2.0 +2.6	+2.8 0	+2.0 +0.2	−2.6 +0.4	+1.8 0	+2.6 +1.4	−3.2 +0.4	+1.8 0	+3.2 +1.4
9.85–12.41	−0.6 +2.6	+2.0 0	+0.6 −0.6	−1.0 +4.0	+3.0 0	+1.0 −1.0	−1.4 +1.8	+2.0 0	+1.4 +0.2	−2.2 +2.8	+3.0 0	+2.2 +0.2	−2.6 +0.6	+2.0 0	+2.6 +1.4	−3.4 +0.6	+2.0 0	+3.4 +1.4
12.41–15.75	−0.7 +2.9	+2.2 0	+0.7 −0.7	−1.0 +4.5	+3.5 0	+1.0 −1.0	−1.6 +2.0	+2.2 0	+1.6 +0.2	−2.4 +3.3	+3.5 0	+2.4 +0.2	−3.0 +0.6	+2.2 0	+3.0 +1.6	−3.8 +0.6	+2.2 0	+3.8 +1.6
15.75–19.69	−0.8 +3.3	+2.5 0	+0.8 −0.8	−1.2 +5.2	+4.0 0	+1.2 −1.2	−1.8 +2.3	+2.5 0	+1.8 +0.2	−2.7 +3.8	+4.0 0	+2.7 +0.2	−3.4 +0.7	+2.5 0	+3.4 +1.8	−4.3 +0.7	+2.5 0	+4.3 +1.8

All data above heavy lines are in accord with ABC agreements. Symbols H7, js6, etc. are hole and shaft designations in ABC system.

*Pairs of values shown represent maximum amount of interference (−) and maximum amount of clearance (+) resulting from application of standard tolerance limits.

Table 10-6. ANSI Standard Interference Locational Fits
(ANSI B4. I-1967, Revised 1974)

Nominal Size Range, Inches	Class LN 1			Class LN 2			Class LN 3		
	Limits of Inter-ference*	Standard Limits		Limits of Inter-ference*	Standard Limits		Limits of Inter-ference*	Standard Limits	
		Hole H6	Shaft n5		Hole H7	Shaft p6		Hole H7	Shaft r6
Over To	Values shown below are given in thousandths of an inch								
0–0.12	0 0.45	+0.25 0	+0.45 +0.25	0 0.65	+0.4 0	+0.65 +0.4	0.1 0.75	+0.4 0	+0.75 +0.5
0.12–0.24	0 0.5	+0.3 0	+0.5 +0.3	0 0.8	+0.5 0	+0.8 +0.5	0.1 0.9	+0.5 0	+0.9 +0.6
0.24–0.40	0 0.65	+0.4 0	+0.65 +0.4	0 1.0	+0.6 0	+1.0 +0.6	0.2 1.2	+0.6 0	+1.2 +0.8
0.40–0.71	0 0.8	+0.4 0	+0.8 +0.4	0 1.1	+0.7 0	+1.1 +0.7	0.3 1.4	+0.7 0	+1.4 +1.0
0.71–1.19	0 1.0	+0.5 0	+1.0 +0.5	0 1.3	+0.8 0	+1.3 +0.8	0.4 1.7	+0.8 0	+1.7 +1.2
1.19–1.97	0 1.1	+0.6 0	+1.1 +0.6	0 1.6	+1.0 0	+1.6 +1.0	0.4 2.0	+1.0 0	+2.0 +1.4
1.97–3.15	0.1 1.3	+0.7 0	+1.3 +0.8	0.2 2.1	+1.2 0	+2.1 +1.4	0.4 2.3	+1.2 0	+2.3 +1.6
3.15–4.73	0.1 1.6	+0.9 0	+1.6 +1.0	0.2 2.5	+1.4 0	+2.5 +1.6	0.6 2.9	+1.4 0	+2.9 +2.0
4.73–7.09	0.2 1.9	+1.0 0	+1.9 +1.2	0.2 2.8	+1.6 0	+2.8 +1.8	0.9 3.5	+1.6 0	+3.4 +2.5
7.09–9.85	0.2 2.2	+1.2 0	+2.2 +1.4	0.2 3.2	+1.8 0	+3.2 +2.0	1.2 4.2	+1.8 0	+4.2 +3.0
9.85–12.41	0.2 2.3	+1.2 0	+2.3 +1.4	0.2 3.4	+2.0 0	+3.4 +2.2	1.5 4.7	+2.0 0	+4.7 +3.5
12.41–15.75	0.2 2.6	+1.4 0	+2.6 +1.6	0.3 3.9	+2.2 0	+3.9 +2.5	2.3 5.9	+2.2 0	+5.9 +4.5
15.75–19.69	0.2 2.8	+1.6 0	+2.8 +1.8	0.3 4.4	+2.5 0	+4.4 +2.8	2.5 6.6	+2.5 0	+6.6 +5.0

All data in this table are in accordance with American-British-Canadian (ABC) agreements.

Limits for sizes above 19.69 inches are not covered by ABC agreements but are given in the ANSI Standard.

Symbols H7, p6, etc. are hole and shaft designations in ABC system.

*Pairs of values shown represent minimum and maximum amounts of interference resulting from application of standard tolerance limits.

that the distances between them be calculated. To determine these distances, the following calculations are necessary:

$$AB = 6 \text{ inches } (9-3)$$
$$AD = 2 \text{ inches } (5-3)$$
$$BD = 4 \text{ inches } (9-5)$$
$$DC = 6 \text{ inches } (8-2)$$

and in the triangles ADC and DBC:

282

Table 10-7. ANSI Standard Force and Shrink Fits (ANSI B4. I-1967, Revised 1974)

Values shown below are in thousandths of an inch

Nominal Size Range, Inches Over — To	Class FN 1 Inter-ference*	Class FN 1 Hole H6	Class FN 1 Shaft	Class FN 2 Inter-ference*	Class FN 2 Hole H7	Class FN 2 Shaft s6	Class FN 3 Inter-ference*	Class FN 3 Hole H7	Class FN 3 Shaft t6	Class FN 4 Inter-ference*	Class FN 4 Hole H7	Class FN 4 Shaft u6	Class FN 5 Inter-ference*	Class FN 5 Hole H8	Class FN 5 Shaft x7
0–0.12	0.05 / 0.5	+0.25 / 0	+0.5 / +0.3	0.2 / 0.85	+0.4 / 0	+0.85 / +0.6				0.3 / 0.95	+0.4 / 0	+0.95 / +0.7	0.3 / 1.3	+0.6 / 0	+1.3 / +0.9
0.12–0.24	0.1 / 0.6	+0.3 / 0	+0.6 / +0.4	0.2 / 1.0	+0.5 / 0	+1.0 / +0.7				0.4 / 1.2	+0.5 / 0	+1.2 / +0.9	0.5 / 1.7	+0.7 / 0	+1.7 / +1.2
0.24–0.40	0.1 / 0.75	+0.4 / 0	+0.75 / +0.5	0.4 / 1.4	+0.6 / 0	+1.4 / +1.0				0.6 / 1.6	+0.6 / 0	+1.6 / +1.2	0.5 / 2.0	+0.9 / 0	+2.0 / +1.4
0.40–0.56	0.1 / 0.8	+0.4 / 0	+0.8 / +0.5	0.5 / 1.6	+0.7 / 0	+1.6 / +1.2				0.7 / 1.8	+0.7 / 0	+1.8 / +1.4	0.6 / 2.3	+1.0 / 0	+2.3 / +1.6
0.56–0.71	0.2 / 0.9	+0.4 / 0	+0.9 / +0.6	0.5 / 1.6	+0.7 / 0	+1.6 / +1.2				0.7 / 1.8	+0.7 / 0	+1.8 / +1.4	0.8 / 2.5	+1.0 / 0	+2.5 / +1.8
0.71–0.95	0.2 / 1.1	+0.5 / 0	+1.1 / +0.7	0.6 / 1.9	+0.8 / 0	+1.9 / +1.4				0.8 / 2.1	+0.8 / 0	+2.1 / +1.6	1.0 / 3.0	+1.2 / 0	+3.0 / +2.2
0.95–1.19	0.3 / 1.2	+0.5 / 0	+1.2 / +0.8	0.6 / 1.9	+0.8 / 0	+1.9 / +1.4	0.8 / 2.1	+0.8 / 0	+2.1 / +1.6	1.0 / 2.3	+0.8 / 0	+2.3 / +1.8	1.3 / 3.3	+1.2 / 0	+3.3 / +2.5
1.19–1.58	0.3 / 1.3	+0.6 / 0	+1.3 / +0.9	0.8 / 2.4	+1.0 / 0	+2.4 / +1.8	1.0 / 2.6	+1.0 / 0	+2.6 / +2.0	1.5 / 3.1	+1.0 / 0	+3.1 / +2.5	1.4 / 4.0	+1.6 / 0	+4.0 / +3.0
1.58–1.97	0.4 / 1.4	+0.6 / 0	+1.4 / +1.0	0.8 / 2.4	+1.0 / 0	+2.4 / +1.8	1.2 / 2.8	+1.0 / 0	+2.8 / +2.2	1.8 / 3.4	+1.0 / 0	+3.4 / +2.8	2.4 / 5.0	+1.6 / 0	+5.0 / +4.0
1.97–2.56	0.6 / 1.8	+0.7 / 0	+1.8 / +1.3	0.8 / 2.7	+1.2 / 0	+2.7 / +2.0	1.3 / 3.2	+1.2 / 0	+3.2 / +2.5	2.3 / 4.2	+1.2 / 0	+4.2 / +3.5	3.2 / 6.2	+1.8 / 0	+6.2 / +5.0
2.56–3.15	0.7 / 1.9	+0.7 / 0	+1.9 / +1.4	1.0 / 2.9	+1.2 / 0	+2.9 / +2.2	1.8 / 3.7	+1.2 / 0	+3.7 / +3.0	2.8 / 4.7	+1.2 / 0	+4.7 / +4.0	4.2 / 7.2	+1.8 / 0	+7.2 / +6.0
3.15–3.94	0.9 / 2.4	+0.9 / 0	+2.4 / +1.8	1.4 / 3.7	+1.4 / 0	+3.7 / +2.8	2.1 / 4.4	+1.4 / 0	+4.4 / +3.5	3.6 / 5.9	+1.4 / 0	+5.9 / +5.0	4.8 / 8.4	+2.2 / 0	+8.4 / +7.0
3.94–4.73	1.1 / 2.6	+0.9 / 0	+2.6 / +2.0	1.6 / 3.9	+1.4 / 0	+3.9 / +3.0	2.6 / 4.9	+1.4 / 0	+4.9 / +4.0	4.6 / 6.9	+1.4 / 0	+6.9 / +6.0	5.8 / 9.4	+2.2 / 0	+9.4 / +8.0

See footnotes at end of table.

283

Table 10-7. ANSI Standard Force and Shrink Fits (ANSI B4.1-1967, Revised 1974) (Continued)

Values shown below are in thousandths of an inch

Nominal Size Range, Inches (Over To)	Class FN 1 Inter-ference*	Class FN 1 Hole H6	Class FN 1 Shaft	Class FN 2 Inter-ference*	Class FN 2 Hole H7	Class FN 2 Shaft s6	Class FN 3 Inter-ference*	Class FN 3 Hole H7	Class FN 3 Shaft t6	Class FN 4 Inter-ference*	Class FN 4 Hole H7	Class FN 4 Shaft u6	Class FN 5 Inter-ference*	Class FN 5 Hole H8	Class FN 5 Shaft x7
4.73–5.52	1.2 / 2.9	+1.0 / 0	+2.9 / +2.2	1.9 / 4.5	+1.6 / 0	+4.5 / +3.5	3.4 / 6.0	+1.6 / 0	+6.0 / +5.0	5.4 / 8.0	+1.6 / 0	+8.0 / +7.0	7.5 / 11.6	+2.5 / 0	+11.6 / +10.0
5.52–6.30	1.5 / 3.2	+1.0 / 0	+3.2 / +2.5	2.4 / 5.0	+1.6 / 0	+5.0 / +4.0	3.4 / 6.0	+1.6 / 0	+6.0 / +5.0	5.4 / 8.0	+1.6 / 0	+8.0 / +7.0	9.5 / 13.6	+2.5 / 0	+13.6 / +12.0
6.30–7.09	1.8 / 3.5	+1.0 / 0	+3.5 / +2.8	2.9 / 5.5	+1.6 / 0	+5.5 / +4.5	4.4 / 7.0	+1.6 / 0	+7.0 / +6.0	6.4 / 9.0	+1.6 / 0	+9.0 / +8.0	9.5 / 13.6	+2.5 / 0	+13.6 / +12.0
7.09–7.88	1.8 / 3.8	+1.2 / 0	+3.8 / +3.0	3.2 / 6.2	+1.8 / 0	+6.2 / +5.0	5.2 / 8.2	+1.8 / 0	+8.2 / +7.9	7.2 / 10.2	+1.8 / 0	+10.2 / +9.0	11.2 / 15.8	+2.8 / 0	+15.8 / +14.0
7.88–8.86	2.3 / 4.3	+1.2 / 0	+4.3 / +3.5	3.2 / 6.2	+1.8 / 0	+6.2 / +5.0	5.2 / 8.2	+1.8 / 0	+8.2 / +7.0	8.2 / 10.2	+1.8 / 0	+11.2 / +10.0	13.2 / 17.8	+2.8 / 0	+17.8 / +16.0
8.86–9.85	2.3 / 4.3	+1.2 / 0	+4.3 / +3.5	4.2 / 7.2	+1.8 / 0	+7.2 / +6.0	6.2 / 9.2	+1.8 / 0	+9.2 / +8.0	10.2 / 13.2	+1.8 / 0	+13.2 / +12.0	13.2 / 17.8	+2.8 / 0	+17.8 / +16.0
9.85–11.03	2.8 / 4.9	+1.2 / 0	+4.9 / +4.0	4.0 / 7.2	+2.0 / 0	+7.2 / +6.0	7.0 / 10.2	+2.0 / 0	+10.2 / +9.0	10.0 / 13.2	+2.0 / 0	+13.2 / +12.0	15.0 / 20.0	+3.0 / 0	+20.0 / +18.0
11.03–12.41	2.8 / 4.9	+1.2 / 0	+4.9 / +4.0	5.0 / 8.2	+2.0 / 0	+8.2 / +7.0	7.0 / 10.2	+2.0 / 0	+10.2 / +9.0	12.0 / 15.2	+2.0 / 0	+15.2 / +14.0	17.0 / 22.0	+3.0 / 0	+22.0 / +20.0
12.41–13.98	3.1 / 5.5	+1.4 / 0	+5.5 / +4.5	5.8 / 9.4	+2.2 / 0	+9.4 / +8.0	7.8 / 11.4	+2.2 / 0	+11.4 / +10.0	13.8 / 17.4	+2.2 / 0	+17.4 / +16.0	18.5 / 24.2	+3.5 / 0	+24.2 / +22.0
13.98–15.75	3.6 / 6.1	+1.4 / 0	+6.1 / +5.0	5.8 / 9.4	+2.2 / 0	+9.4 / +8.0	9.8 / 13.4	+2.2 / 0	+13.4 / +12.0	15.8 / 19.4	+2.2 / 0	+19.4 / +18.0	21.5 / 27.2	+3.5 / 0	+27.2 / +25.0
15.75–17.72	4.4 / 7.0	+1.6 / 0	+7.0 / +6.0	6.5 / 10.6	+2.5 / 0	+10.6 / +9.0	9.5 / 13.6	+2.5 / 0	+13.6 / +12.0	17.5 / 21.6	+2.5 / 0	+21.6 / +20.0	24.0 / 30.5	+4.0 / 0	+30.5 / +28.0
17.72–19.69	4.4 / 7.0	+1.6 / 0	+7.0 / +6.0	7.5 / 11.6	+2.5 / 0	+11.6 / +10.0	11.5 / 15.0	+2.5 / 0	+15.6 / +14.0	19.5 / 23.6	+2.5 / 0	+23.6 / +22.0	26.0 / 32.5	+4.0 / 0	+32.5 / +30.0

All data above heavy lines are in accordance with American-British-Canadian (ABC) agreements. Symbols H6, H7, s6, etc. are hole and shaft designations in ABC system. Limits for sizes above 19.69 inches are not covered by ABC agreements but are given in the ANSI Standard.

*Pairs of values shown represent minimum and maximum amounts of interference resulting from application of standard tolerance limits.

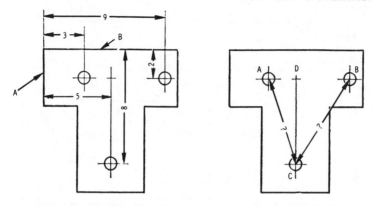

Fig. 10-3. Typical dimension drawing for machine work (left), with duplicate drawing indicating additional dimensions that are necessary for precision work (right).

$$AC = \sqrt{(AD)^2 + (DC)^2} = \sqrt{(2)^2 + (6)^2} = \sqrt{40} = 6.325 \text{ inches}$$
$$BC = \sqrt{(DC)^2 + (DB)^2} = \sqrt{(6)^2 + (4)^2} = \sqrt{52} = 7.211 \text{ inches}$$

Thus, considerable time can be saved for the machinist by placing these dimensions on the drawing.

The method to be employed in laying out holes that are to be drilled depends on the accuracy desired, that is, whether the holes are to register with other holes or fixed studs. If extreme accuracy is not required, the centers for the holes can be laid out with a chalk pencil and steel rule. For locating the holes accurately, as in jig and experimental work, the centers for the holes must be laid out and scribed on the surface of the work. Scribing or layout must be done with precision tools, such as a sharp-pointed scriber, dividers, surface gage, and surface plate.

The machined surfaces should be cleaned and a copper sulfate (blue vitriol) solution applied to the surface. The treated surface, when dry, will distinctly show any lines that are made on it. Chalk can be well rubbed into the surface for the less accurate jobs.

Laying Out the Workpiece

The link (Fig. 10-4) can be used here to serve as an example of a typical workpiece for precision layout. The hole centers A and B

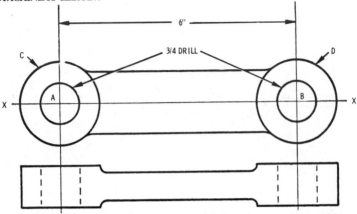

Fig. 10-4. Top and side views of a link, illustrating precision necessary for layout, centering, and boring operations.

are to be made concentrically with respect to the perimeters C and D of the hubs.

With the dividers set to a radius slightly longer than the radius of the hub, describe intersecting arcs from several points along the hub perimeter, as shown in Figs. 10-5 and 10-6. Here the pairs of arcs intersect at points A, B, C, and D, forming the enclosed four-sided figure a b c d, the center point of which is the center point of the hole that is to be drilled. If the layout were made accurately, the hole would be concentrically located with the center point of the hub.

To locate the center point, the two axes AB and CD are scribed, intersecting at center point O. The dividers can be used to check the center point for accuracy; the distances OA', OB', OC', and OD' should all be equal. If the center point should be "out" on any axis, it can be shifted until four equal measurements are obtained. The center point can then be located permanently by making a light mark with the center punch, being careful to hold the punch perpendicular to the surface.

After locating the center points of the hubs of the link, the next step is to scribe the longitudinal axis XX (see Fig. 10-2). This can be done accurately by means of a surface plate and surface gage, as shown in Fig. 10-7. First set the scriber to register with the center point O, and scribe the line X across the left-hand end of the link

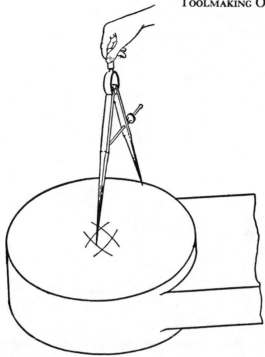

Fig. 10-5. Describing intersecting arcs to locate the center point of the hub of the link so that the hole to be drilled will be concentric with the circumference of the hub.

(position *A*); repeat with the same setting on the opposite end of the link (position *B*). The dividers (or trammels preferably) should be set to exactly 6 inches; with one point on the center point *O*, an arc *ab* is scribed on the opposite end of the link, intersecting the axis *XX* at the center point *O'* which is the correct center point for the right-hand end of the link (Fig. 10-8). The center point *O* should be marked permanently with the center punch.

Drilling Center Holes

After the center points have been laid out, and prior to drilling, circles representing the holes are scribed with dividers. Then the center punch marks should be enlarged to aid in starting the twist drill. Begin to drill the hole at center point *O* (see Fig. 10-8) with the point of the twist drill in the enlarged center punch mark. The

287

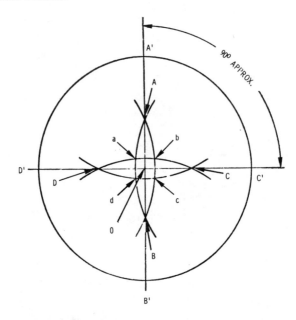

Fig. 10-6. Determining the center point of the figure enclosed by the intersecting arcs to locate the center point of the hub of the link.

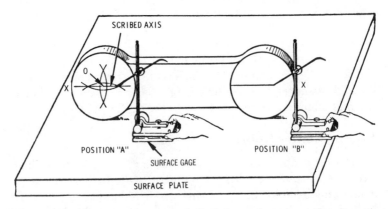

Fig. 10-7. Using the surface plate and the surface gage to scribe the longitudinal axis (XX) of the link.

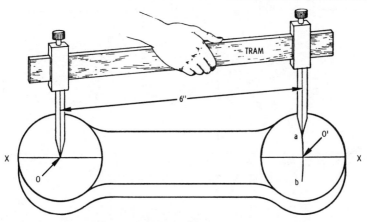

Fig. 10-8. A trammel can be used to scribe the center point (O') of the opposite end of the link.

hand feed should be used until a "dimple" is made in the workpiece. Then the dimple should be examined to determine whether it is "on center" with the scribed circle.

If the drill point or dimple has tended to slide off to one side, a center gouge or center chisel can be used to "draw" the drill point back to the correct position, as shown in Fig. 10-9. First determine the side toward which the drill must be shifted. Then chisel a small groove in that side (see Fig. 10-9). If the dimple is very eccentric, several chiseled grooves can be required; however, if it is only slightly off center, a small groove will correct the position of the drill. As the grooves enable to drill point to cut more easily, it is "drawn" toward the side cut by the gouge or chisel. This operation cannot be done after the full diameter of the cut has been reached.

Locating Center Points with Precision

Methods of locating center points with instruments that depend on the toolmaker's "sense of touch" for accuracy are called contact methods. Examples of these instruments are micrometers, vernier calipers, dial gages, etc. The experienced toolmaker with a finely developed sense of touch and an accurate instrument can obtain a surprising degree of accuracy by the direct-contact measurement method. Accuracy within limits of 0.0001 inch can be obtained by persons who possess a sensitive touch. The following contact

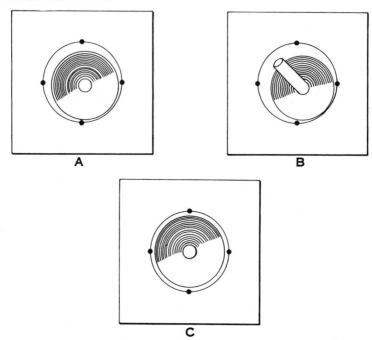

Fig. 10-9. One method used to draw a drill point to bring it true with the center point of the hole, showing three positions as follows: (A) dimple is off center; (B) gouge mark on side toward which the drill point is to be drawn; and (C) dimple has returned to the center point.

methods are used by toolmakers to locate center points with precision: (1) button, (2) disk, (3) disk and button, (4) multidiameter disks, (5) size blocks, and (6) master plate.

Toolmakers' buttons—This method of locating center points is well adapted to small precision work. The buttons are usually 0.3, 0.4, or 0.5 inch in diameter, and they are usually ground and lapped to uniform size; the ends are finished square, or 90° to the cylindrical surface. Preferably, the outside diameter of the buttons should be such that the radius can be determined easily, and the hole through the center should be about ⅛ (0.125) inch larger than the retaining screw, so that the button can be adjusted laterally. The button method of locating center points for drilling holes requires a series of seven operations: (1) laying out centers, (2) marking center points with center punch or prick punch, (3) drilling holes for button screws, (4) tapping holes for button screws, (5) clamping buttons in approximate position, (6)adjusting the buttons to the correct positions, and (7) boring the holes.

A working drawing of a jig plate (Fig. 10-10) can be used to illustrate the button method of locating center points. The dimensions necessary for locating three holes in the jig plate are given in the drawing.

After *laying out* the center points for the holes and *marking* them lightly with a center punch, the holes for the button screws should be *drilled*. Care should be taken to select the correct size of tap drill. A tap drill should be selected that will give a snug fit, rather than the usual commercial loose fit.

Then the holes for the button screws should be *tapped*. A typical button screw is ⅛ (0.125) inch in diameter with 40 threads per inch. Any burrs that project from the button ends after tapping should be removed carefully.

The button screws are used to clamp the buttons in an approxi-

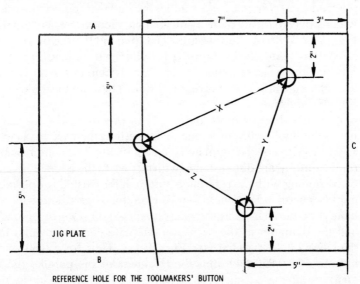

Fig. 10-10. Working drawing for a jig plate. Note that the dimensions, X, Y, and Z are important in the toolmakers' button method of locating center points.

mate position. As the holes in the toolmakers' buttons are larger than the diameter of the button screw, the buttons are not located concentrically with the button screws, with introduces an error (Fig. 10-11). A second error is introduced because the button screws have not been located with precision.

Let the axis AB represent the true center line, as per dimensions on the working drawing (see Fig. 10-10). Regardless of the care exercised in laying out the center points on the jig plate, the intersections of the center lines are not the precise center points of the holes to be drilled.

If we assume that the axis AB is the true center line, the axis CD of the button screw, because of inaccuracies in layout, can be "off" a distance GH, which introduces the *first* error. Also, the button is not necessarily concentric with the button screw, indicated by the axis EF, because of clamping. It may be "off" a distance IJ, which introduces the *second* error, giving an accumulation of errors, or total error BF.

In *adjusting the buttons to their correct positions*, one button is selected as a *reference button* and is located in its correct position with respect to the jig plate by means of a micrometer, as shown in Fig. 10-12. With the button screw clamped lightly, the button can be shifted minutely by rapping it lightly with a lead hammer. Micrometer measurements can be taken at each shift of the button until the button is in its correct position. Then the button can be clamped tightly.

To center two or more buttons with precision, the measurements are made with reference to the buttons themselves, as well as to the edges of the jig plate (see Fig. 10-10). The center points for the buttons are dimensioned with respect to the edges A, B, and C. Beginning with the reference button, it is adjusted for position until its center is 10 inches (7 + 3) from the edge C and 5 inches from the edge A. In using the micrometer to adjust the button, add half the diameter of the button to the plate dimensions plus the thickness of the parallel (see Fig. 10-12). Thus for a 0.500-inch button, add 0.250 inch plus the thickness of the parallel (0.250 inch), to the plate dimension (7 inches). Therefore, to locate the reference button at 10 inches (7 + 3) from the edge C, set the micrometer at 10.500 inches (10 + 0.250 + 0.250), and adjust the button to that distance, as shown in Fig. 10-12.

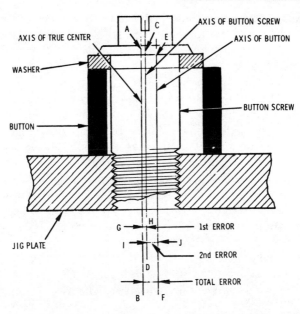

Fig. 10-11. An enlarged view of a toolmakers' button, showing the errors introduced and the necessity of tapping adjustments to correct them.

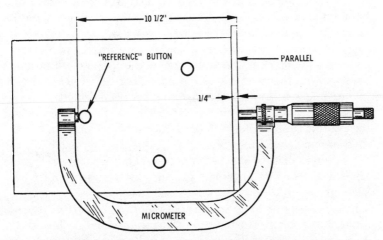

Fig. 10-12. Using a micrometer and a parallel to locate the "reference" button on a jig plate.

293

The next step is to adjust the button with respect to the edge A. Similarly, locate the other buttons in their correct positions. Then clamp the reference button tightly, leaving the other two buttons clamped lightly.

The final precision adjustments involve locating the buttons in their correct positions to correspond with the measurements X, Y, and Z. In making these adjustments, the micrometer reading depends on whether an outside or an inside micrometer is used. For outside measurements, the diameter of the button (0.500 inch) must be added to the dimensions (X, Y, and Z), and subtracted if an inside measurement is used, as shown in Fig. 10-13. In making these adjustments, it can be noted that the reference button (which was clamped tightly) is not moved, but the other buttons (clamped lightly) are shifted to their correct positions and clamped tightly (Fig. 10-14).

After the toolmakers' buttons have been positioned correctly and clamped tightly, the final step in locating the center points of the holes for drilling or boring to size can be taken. The jig plate is then mounted on the faceplate of a lathe, with the button approximately "on center" with the axis of rotation, clamping the jig plate lightly (Fig. 10-15). Rotate the faceplate by hand, checking the button with a dial indicator. Adjust the jig plate by rapping lightly with a lead hammer until the hand of the dial ceases to move, and tighten the clamping bolts. Then remove the toolmakers' button and drill or bore to the correct size.

When the workpiece is heavy and most of the jig plate is "off center," a counterweight can be attached to prevent vibration in drilling. The other toolmakers' buttons can be centered and drilled in the same manner after the first hole has been drilled.

Disk—This method is similar in principle to the button method for locating center points. The disk method is different in that the disks are made with the dimensions of the diameters such that the center points of the disks are in the correct position for drilling the holes when the circumferences of the disks are in contact with each other.

If three holes are to be drilled at distances X, Y, and Z from each other, the problem is to determine the required size of three disks, A, B, and C, so that their center points will be located at the points a, b, and c, when the circumferences of the disks·are in contact

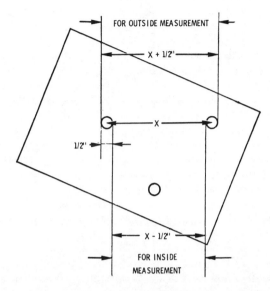

Fig. 10-13. Diagram showing corrections necessary in setting inside and outside micrometers.

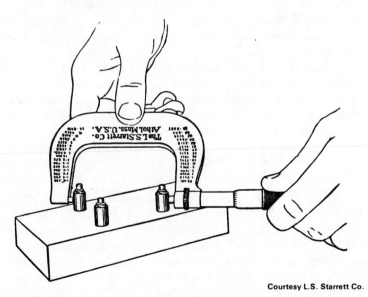

Fig. 10-14. Using the micrometer to adjust the toolmakers' buttons.

295

(Fig. 10-16). The following steps can be taken to locate the center points:

1. First subtract the dimension *Y* from dimension *X*. This results in the distance $a'b$, which is the difference between the radii of the disks B and C.
2. Add the distance $a\ b$ to the dimension Z, which will give the diameter of disk B. Describe the circumference of disk B, using the point b as the center point.

To accomplish this, rotate the point a' around the center point b to the point b'; then the distance $b'c$ is equal to the diameter of disk B. As the radius of disk B is needed, it can be obtained by bisecting the line $b'c$, using the arcs m and n. Using the line oc as the radius and the point b as center point, describe the circumference of the disk B.

3. Describe the disk C tangent to disk B.
4. Describe disk A so that it will be tangent to both disk B and disk C.

It can be noted in the illustration (see Fig. 10-16) that when the three disks A, B, and C are tangent to each other, the center points of the disks coincide with the center points a, b, and c at the precise dimensions *X*, *Y*, and *Z*.

To locate the center points for holes that are to be drilled, the finished disks can be fastened to the workpiece in their correct locations with their circumferences in contact. As in the button method, the workpiece can be mounted on a faceplate in the lathe (see Fig. 10-15) and one disk centered by means of the dial indicator applied to the center point of the disk. After the first disk has been centered, it can be removed and the hole drilled or bored. The same procedure can then be followed for the remaining holes.

Disks and buttons—Toolmakers' buttons and disks can be used to locate center points in a combination method that combines the principles of the two methods described previously. Each disk must have a hole at its center that fits the toolmakers' button accurately. Also, a bushing having the same diameter as that of the button and a hole in which a center punch can slide should be provided.

A diagram illustrating the disk-and-button method of locating

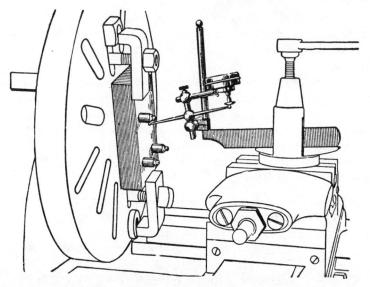

Fig. 10-15. Using the dial indicator to center the toolmakers' button on the faceplate of a lathe.

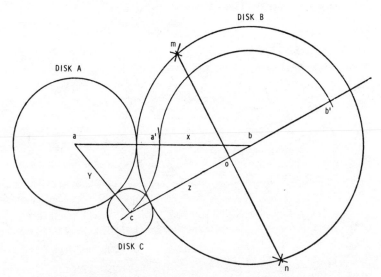

Fig. 10-16. Diagram showing the disk method of locating the center points of holes by means of different-sized disks.

center points of holes is shown in Fig. 10-17. In reference to the diagram, the problem is to locate the center points of six holes equally spaced on the circumference of a circle that is 6 inches in diameter.

The size of the disks must be determined first. The size of the smaller disks can be determined by the following rule. *Rule:* Multiply the diameter of the circle on which the center points of the disks are located by the sine of one-half the number of degrees in the angle between the adjacent disks.

Applying the rule to the diagram (see Fig. 10-17), the problem can be solved as follows:

$$\text{Angle between centers of adjacent disks} = 360° \div 6 = 60°$$
$$\text{One-half the angle between centers} = 60° \div 2 = 30°$$
$$\text{Sine of } 30° = 0.500$$
$$\text{Diameter of the smaller disks} = 6 \times 0.500 = 3 \text{ inches}$$
$$\text{Diameter of the central disk} = 6 - 3 = 3 \text{ inches}$$

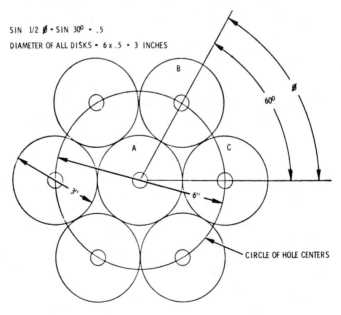

Fig. 10-17. Diagram showing the disk-and-button method of locating the center points of holes.

Thus, in this instance, all the disks are the same size as shown in Fig. 10-17. The disks and toolmakers' buttons can then be mounted on the workpiece, as explained in the previous methods.

Measuring angles—Measuring angles between two or more lines or surfaces ("reading the angles") can be accomplished with a variety of tools, depending upon the accuracy required for the job. For simple angles, a common protractor, as shown in Fig. 10-18, might serve. These are graduated in degrees from 0 to 180°. The scale on this model can also be used as a depth gage.

A bevel protractor with a vernier attachment such as the one in Fig. 10-19 can accurately measure angles to 5 minutes or one-twelfth of a degree. Fig. 10-20 shows this bevel protractor in use checking the angle of a surface.

Multidiameter disks for angular location—A line can be located

Courtesy L.S. Starrett Co.

Fig. 10-18. A protractor.

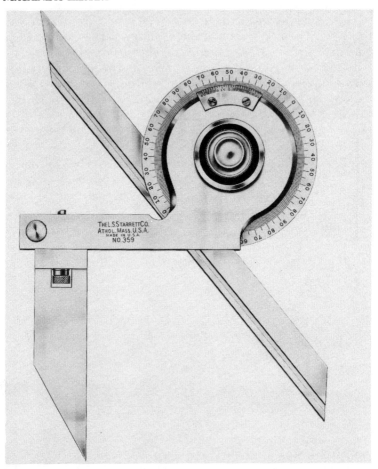

Fig. 10-19. Universal bevel protractor.

at a given angle with another line by means of two disks of different diameters located with the correct distance between their center points. This method can be used to machine a surface at a given angle with another surface that is already finished.

A diagram illustrating angular location by means of *nontangent disks* is shown in Fig. 10-21. In reference to the diagram, if the lower edge of the workpiece is finished and the upper edge is to be

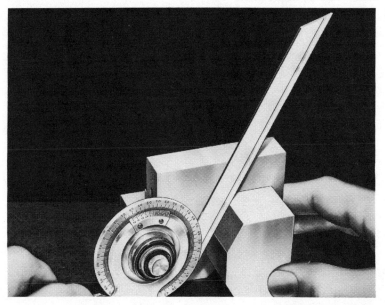

Fig. 10-20. Checking the angle of a surface with a universal bevel protractor.

milled at an angle of 20°, the problem is to determine the proper distance *AB* to place two disks (1 inch and 2 inches in diameter) so that the upper edge of the workpiece will be tangent to both disks at the given angle. A rule can be used to solve the problem. *Rule:* Divide the difference between the radii of the two disks by the sine of half the required angle.

Applying the rule to the diagram (see Fig. 10-21), the correct distance *AB* between center points of the disks can be determined as follows:

$$
\begin{aligned}
\text{Difference between radii} &= 1 \text{ inch } - 0.500 \text{ inch} \\
&= 0.500 \text{ inch} \\
\text{Sine } \tfrac{1}{2}\,\theta &= \sin 10° \\
&= 0.1736 \\
\text{Distance between center points } A \text{ and } B &= \tfrac{1}{2} \div 0.1736 \\
&= 2.88 \text{ inches}
\end{aligned}
$$

Thus, the center points of the two disks *A* and *B* (see Fig. 10-21) can be placed at a distance *AB* of 2.88 inches to mill the upper edge

301

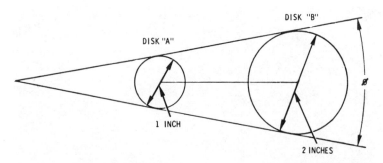

Fig. 10-21. Diagram showing the angular location of a given line by means of nontangent disks.

of the workpiece at an angle of 20° with the finished lower edge.

A diagram illustrating angular location by means of *tangent disks* is shown in Fig. 10-22. In reference to the diagram, the problem is to determine the relative sizes of two tangent disks, rather than the distance between center points, so that lines that are common tangents to the disks will be located at the desired angle.

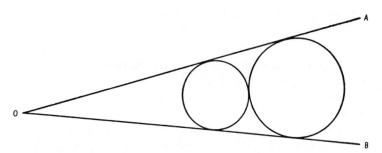

Fig. 10-22. Diagram showing the angular location of a given line by means of tangent disks.

It can be noted in the diagram (see Fig. 10-22) that the inclination of the line *OA* depends on the relative sizes of the two disks. The angle *AOB* can be made either larger or smaller by either increasing or decreasing the diameter of the larger disk. For example, if the required angle *AOB* is 20° and the diameter of the smaller disk is 1 inch, the diameter of the larger disk can be found by the application of the following three rules:

302

1. Multiply twice the diameter of the smaller disk by the sine of ½ the required angle (sin ½ θ or sin 10° = 0.1736):

$$2 \times 1 \times 0.1736 = 0.3473$$

2. Divide the product by *1* minus the sine of ½ the required angle:

$$0.3473 \div (1 - 0.1736) = 0.42025$$

3. Add the quotient to the diameter of the smaller disk to obtain the diameter of the larger disk:

$$1 + 0.42025 = 1.42025 \text{ inches}$$

Thus, 1.42025 inches is the required diameter of the larger disk to form the required angle *AOB* of 20°, as shown in Fig. 10-22. This method is accurate for obtaining any angle of inclination up to 40°.

A diagram illustrating angular location by means of a *disk and combination square* is shown in Fig. 10-23. In reference to the diagram, the problem is to determine the proper setting (distance *X*) on the blade for the head of the square to give the required angle. For example, if the required angle is 20°, and a disk with a 2-inch diameter is to be used, the setting (distance *X*) on the blade of the square can be found as follows:

1. Multiply the radius of the disk by the cotangent of ½ the desired angle—cot ½ θ or cot 10° = tan(90° − 10°) = tan 80° = 5.6713:

$$1 \times 5.6713 = 5.6713$$

2. Add to this product the radius of the disk:

$$5.6713 + 1 = 6.6713 \text{ inches}$$

Thus, the proper setting (distance *X* in Fig. 10-23) on the blade of the combination square should be 6.6713 to give the required 20° angle.

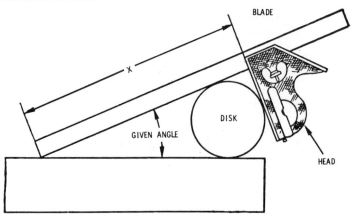

Fig. 10-23. Diagram showing angular location of a given line by means of a disk and combination square.

Size blocks—A size block is a rectangular piece that is machined to precise dimensions. Size blocks are available in various widths, so that either a single block or a combination of size blocks can be selected to locate the work properly. Size blocks provide a convenient method for locating workpieces so that holes can be drilled or bored at required center-to-center distances.

The jig plate (Fig. 10-24) is an example of a workpiece on which size blocks can be used to locate the center points *A* and *B* for two holes that are to be drilled. We can assume that the jig plate is to be machined to a rectangular shape, and that the center points *A* and *B* for the two holes are located as the dimensions require.

The first operation is to fasten two parallels to a faceplate of the lathe at 90° to each other, as shown in Fig. 10-25. The parallels should be set off from the center of rotation at the distances given by the dimensions in the dimensional drawing (see Fig. 10-24); that is, one parallel is offset 4 inches, and the other parallel is offset 6 inches. The accuracy of locating the holes depends on the precision with which the parallels are located. If the parallels are not offset precisely at 4 inches and 6 inches from center, the holes will not be correctly located, of course. Another source of error is lack of preciseness in placing the parallels at 90° to each other.

After attaching the parallels to the faceplate securely, the next step is to mount the workpiece on the faceplate so that it is in

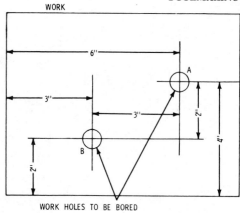

Fig. 10-24. Dimensional drawing of a jig plate, used as an example of a workpiece on which size blocks can be used to locate the center points for holes that are to be drilled.

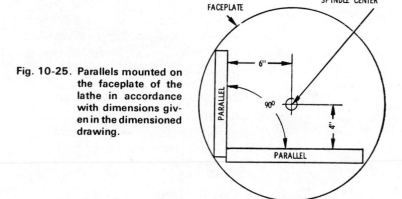

Fig. 10-25. Parallels mounted on the faceplate of the lathe in accordance with dimensions given in the dimensioned drawing.

contact with both parallels, as shown in Fig. 10-24. This places the workpiece correctly for drilling hole A. Then the workpiece can be shifted to the correct position for drilling the second hole B by inserting size blocks between the workpiece and the parallels (see Fig. 10-26).

To determine the correct size blocks that can be used, note in Fig. 10-26 that the center point B is located a distance of 2 inches

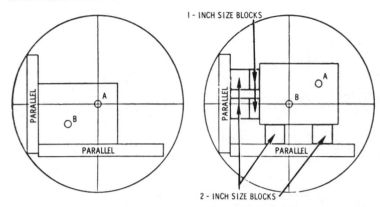

1 - INCH SIZE BLOCKS

2 - INCH SIZE BLOCKS

Fig. 10-26. Workpiece positioned by means of parallels for drilling the first hole (left), and positioned by means of parallels and size blocks for drilling the second hole (right).

from the edge of the workpiece; therefore, 2-inch blocks can be used on that side. Likewise, the center point *B* is offset a distance of 3 inches from the center point *A*. Assuming that both 1-inch and 2-inch size blocks are available but 3-inch size blocks are not available the workpiece can be shifted a total of 3 inches from the parallel by combining the 1-inch and 2-inch size blocks (see Fig. 10-26), to position the workpiece correctly for drilling the second hole *B*.

Master plate—This device is designed to be mounted on the faceplate of a lathe; corresponding holes in the master plate and in the workpiece enable the workpiece to be shifted to correct position for drilling these holes in the workpiece. Master plates are made in any number of designs to accommodate workpieces of various shapes.

A dimensioned drawing of a jig plate is shown in Fig. 10-27 as an example of a workpiece for which a master plate would be suitable for locating the center points for holes to be drilled or bored. The dimensioned drawing shows the location of holes that are to be drilled in the workpiece; the master plate must have corresponding holes so that the workpiece can be shifted into correct position for drilling the holes, as shown in Fig. 10-28.

A step is provided on each side of the master plate (see Fig. 10-28) for clamping, and a projecting parallel fixes the lateral

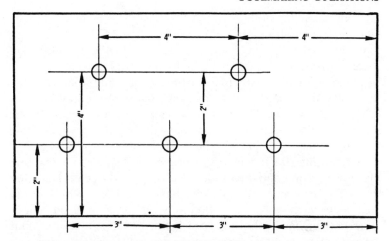

Fig. 10-27. Dimensioned drawing of a jig plate, which is suitable for using as a master plate to locate the center points of holes that are to be drilled in the plate.

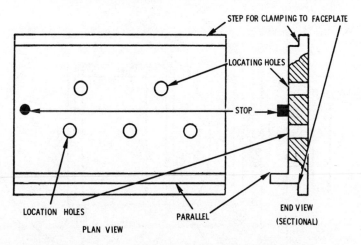

Fig. 10-28. A master plate designed for positioning the jig plate.

position of the work. A stop at one end of the master plate accurately positions the workpiece longitudinally. The work is positioned against both the parallel and the stop for drilling the holes. The holes are always at the correct position from the edges of the workpiece regardless of the number of duplicates made.

307

Clamps and tap bolts or screws that pass through the workpiece and into the master plate can be used to mount the workpiece on the master plate. If these cannot be used conveniently, solder can be used for fastening the work to the master plate.

In mounting the master plate on the faceplate of the lathe, one of the holes of the master plate is engaged with the central locating plug (Fig. 10-29), and the master plate is then clamped to the faceplate. The setup is then ready for drilling or boring the hole corresponding to the hole in the master plate.

After drilling the first hole, the master plate is shifted so that first one locating hole and then another can engage the central locating plug. To locate the center of a hole that is to be bored with extreme accuracy, the hole is first drilled to within 0.005 or 0.006 inch of correct size; then it is bored to nearly finished size, a small amount of material remaining for reaming or grinding to finished size. The holes are not drilled to finished size because the point of the twist drill tends to wander off the center point.

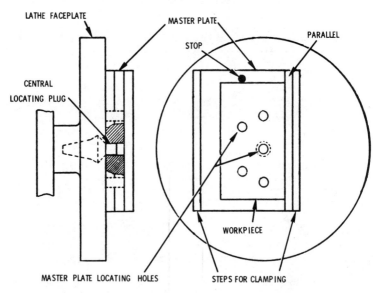

Fig. 10-29. Side view (left) showing the master plate mounted on the face-plate of the lathe; and front view (right) showing the workpiece mounted on the master plate, which is in position for drilling a hole.

Checking the Square

The four-disk method can be used to check a square for trueness of the 90° angle (Fig. 10-30). Four disks of the same size are required. The accuracy of the check depends on the disks being precisely the same size.

The disks are placed in tangency with themselves and the square (see Fig. 10-30). Either outside calipers or inside calipers can be used for measuring. Using outside calipers, measure the distances *AB* and *CD*. If these distances are equal, the angle between the beam and the blade of the square is true at 90°. Similarly, using inside calipers, measure the distances *EF* and *GH*. If these distances are equal, the square accurately indicates 90°.

Sine Bar for Measuring Angles

A *sine bar* is a straightedge with two attached cylindrical plugs, as shown in Fig. 10-31. The instrument is used to measure angles.

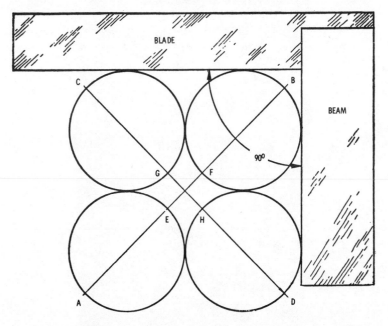

Fig. 10-30. The four-disk method of checking a square for trueness of the 90° angle.

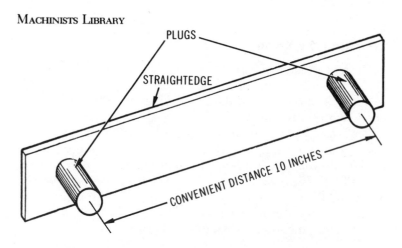

Fig. 10-31. A sine bar. It is used for making angular measurements.

The fundamental principle of the sine bar is illustrated in Fig. 10-32. The plugs are placed at a convenient distance (10 inches) on the straightedge. The imaginary line joining the center points of the two plugs represents the hypotenuse of a right triangle, as shown in the illustration.

As noted in Fig. 10-32:

$$Sin = BC \div AC$$

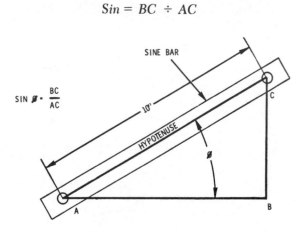

Fig. 10-32. Diagram showing the fundamental principle of the sine bar.

As the hypotenuse *AC* is known (10 inches), it is necessary only to measure the length of the side *BC* and divide by the length of the hypotenuse *AC* to give the sine of the angle *BAC*. The sine of the angle can be derived from the sine table (Table 10-8).

As it is inconvenient to measure the distance *BC* from the center point of the plug, a more convenient method can be used to obtain this measurement (Fig. 10-33). The sine bar can be placed in position above a reference surface, such as a surface plate, and the distances *A* and *B* measured accurately with a micrometer. The distance *A* can be subtracted from the distance *B* to determine the distance (*B A* in Fig. 10-33), which corresponds to the measurement *BC* in Fig. 10-32.

For example, if the height *A* = 2.14 inches and *B* = 7.256 inches (see Fig. 10-33), the angle of inclination of the sine bar can be determined as follows:

$$\text{Sin } \theta = (7.256 - 2.14) \div 10 = 0.5116$$

by interpolating from the table of sines

$$\text{Angle } \theta = 30°46' \quad \text{or} \quad 30.77046383°$$

Fig. 10-34 shows a sine bar set up to check the angle of a gage

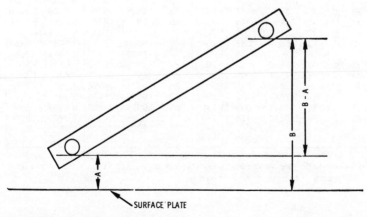

Fig. 10-33. Diagram showing the use of the sine bar to obtain the measurement BC in Fig. 10-32.

311

Table 10-8. Natural Trigonometrical Functions

Degree	Sine	Cosine	Tangent	Secant	Degree	Sine	Cosine	Tangent	Secant
0	0.00000	1.0000	0.00000	1.0000	46	0.7193	0.6947	1.0355	1.4395
1	0.01745	0.9998	0.01745	1.0001	47	0.7314	0.6820	1.0724	1.4663
2	0.03490	0.9994	0.03492	1.0006	48	0.7431	0.6691	1.1106	1.4945
3	0.05234	0.9986	0.05241	1.0014	49	0.7547	0.6561	1.1504	1.5242
4	0.06976	0.9976	0.06993	1.0024	50	0.7660	0.6428	1.1918	1.5557
5	0.08716	0.9962	0.08749	1.0038	51	0.7771	0.6293	1.2349	1.5890
6	0.10453	0.9945	0.10510	1.0055	52	0.7880	0.6157	1.2799	1.6243
7	0.12187	0.9925	0.12778	1.0075	53	0.7986	0.6018	1.3270	1.6616
8	0.1392	0.9903	0.1405	1.0098	54	0.8090	0.5878	1.3764	1.7013
9	0.1564	0.9877	0.1584	1.0125	55	0.8192	0.5736	1.4281	1.7434
10	0.1736	0.9848	0.1763	1.0154	56	0.8290	0.5592	1.4826	1.7883
11	0.1908	0.9816	0.1944	1.0187	57	0.8387	0.5446	1.5399	1.8361
12	0.2079	0.9781	0.2126	1.0223	58	0.8480	0.5299	1.6003	1.8871
13	0.2250	0.9744	0.2309	1.0263	59	0.8572	0.5150	1.6643	1.9416
14	0.2419	0.9703	0.2493	1.0306	60	0.8660	0.5000	1.7321	2.0000
15	0.2588	0.9659	0.2679	1.0353	61	0.8746	0.4848	1.8040	2.0627
16	0.2756	0.9613	0.2867	1.0403	62	0.8820	0.4695	1.8807	2.1300
17	0.2924	0.9563	0.3057	1.0457	63	0.8910	0.4540	1.9626	2.2027
18	0.3090	0.9511	0.3249	1.0515	64	0.8988	0.4384	2.0503	2.2812
19	0.3256	0.9455	0.3443	1.0576	65	0.9063	0.4226	2.1445	2.3662
20	0.3420	0.9397	0.3640	1.0642	66	0.9135	0.4067	2.2460	2.4586
21	0.3584	0.9336	0.3839	1.0711	67	0.9205	0.3907	2.3559	2.5593
22	0.3746	0.9272	0.4040	1.0785	68	0.9272	0.3746	2.4751	2.6695
23	0.3907	0.9205	0.4245	1.0864	69	0.9336	0.3584	2.6051	2.7904
24	0.4067	0.9135	0.4452	1.0940	70	0.9397	0.3420	2.7475	2.9238
25	0.7880	0.6157	1.2799	1.6243	71	0.9455	0.3256	2.9042	3.0715
26	0.4385	0.8988	0.4877	1.1126	72	0.9511	0.3090	3.0777	3.2361
27	0.4540	0.8910	0.5095	1.1223	73	0.9563	0.2924	3.2709	3.4203
28	0.4695	0.8829	0.5317	1.1326	74	0.9613	0.2756	3.4874	3.6279
29	0.4848	0.8746	0.5543	1.1433	75	0.9659	0.2588	3.7321	3.8637
30	0.5000	0.8660	0.5774	1.1547	76	0.9703	0.2419	4.0108	4.1336
31	0.5150	0.8572	0.6009	1.1666	77	0.9744	0.2250	4.3315	4.4454
32	0.5299	0.8480	0.6249	1.1792	78	0.9781	0.2079	4.7046	4.8007
33	0.5446	0.8387	0.6494	1.1924	79	0.9816	0.1908	5.1446	5.2408
34	0.5592	0.8290	0.6745	1.2062	80	0.9848	0.1736	5.6713	5.7588
35	0.5736	0.8192	0.7002	1.2208	81	0.9877	0.1564	6.3138	6.3924
36	0.5878	0.8090	0.7265	1.2361	82	0.9903	0.1392	7.1154	7.1853
37	0.6018	0.7986	0.7536	1.2521	83	0.9925	0.12187	8.1443	8.2055
38	0.6157	0.7880	0.7813	1.2690	84	0.9945	0.10453	9.5144	9.5668
39	0.6293	0.7771	0.8098	1.2867	85	0.9962	0.08716	11.4301	11.474
40	0.6428	0.7660	0.8391	1.3054	86	0.9976	0.06976	14.3007	14.335
41	0.6561	0.7547	0.8693	1.3250	87	0.9986	0.05234	19.0811	19.107
42	0.6691	0.7431	0.9004	1.3456	88	0.9994	0.03490	28.6363	28.654
43	0.6820	0.7314	0.9325	1.3673	89	0.9998	0.01745	57.2900	57.299
44	0.6947	0.7193	0.9657	1.3902	90	1.0000	Inf.	Inf.	Inf
45	0.7071	0.7071	1.0000	1.4142					

NOTE—For intermediate values reduce angles from degrees, minutes and seconds to degrees and decimal parts of a degree (as 40° 21' 30" = 40.358°) interpolate or consult a larger table.

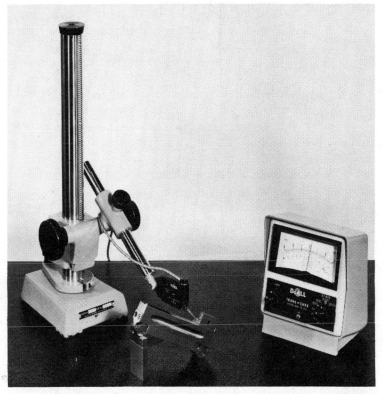

Courtesy DoAll Co.

Fig. 10-34. Sine bar and electronic height gage being used to check an angle.

block end standard. The plugs of the sine bar are attached at the ends of the bar so one is in contact with the surface plate. The gage blocks supporting the other plug are the exact height required. In this case, an electronic height gage is being used to check the angle. This electronic height gage is set to measure on the 0.00001-inch scale.

Sine bars can be made for special functions. Fig. 10-35 shows a sine bar that has a head and tailstock to hold work between centers when checking tapers. A precision ground shaft is being held between centers, and the taper is being checked with an electronic height gage.

313

Sine plates (see Fig. 10-36) are also used extensively in industry. They are based upon the same principle as the sine bar but have a wider surface for holding the work to be checked. Some are used to hold work while being machined, especially during grinding operations.

Note that in these illustrations of sine bars and plates gage blocks are being used to support one end of the sine bar or plate. These are accurate within a few millionths of an inch and come in sets, such as shown in Fig. 10-37, that can measure in steps as small as 0.00005 inch. The illustration shows how five of these blocks can be combined to equal 2.5214 inch. Such a set can usually measure up to 12 inches.

Another method of extremely accurate angular measurement is the use of angle gage blocks. These are similar to gage blocks and provide angles as close as one-fourth of a second of a degree. A set

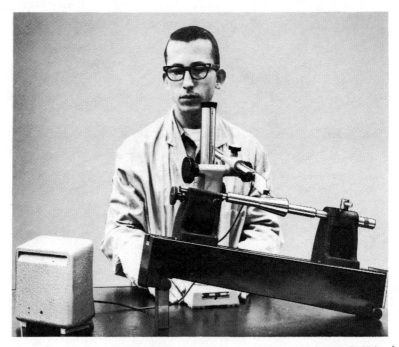

Courtesy DoAll Co.

Fig. 10-35. Sine bar with head and tailstock.

314

Fig. 10-36. Sine plate in use.

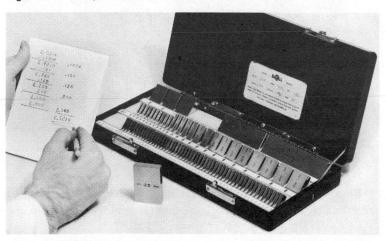

Fig. 10-37. A set of gage blocks.

315

of only 16 gage blocks, as seen in Fig. 10-38, can be used to measure angles from 0° to 99° in steps of one second. These have the advantage of being assembled easily and quickly while requiring only simple addition and subtraction.

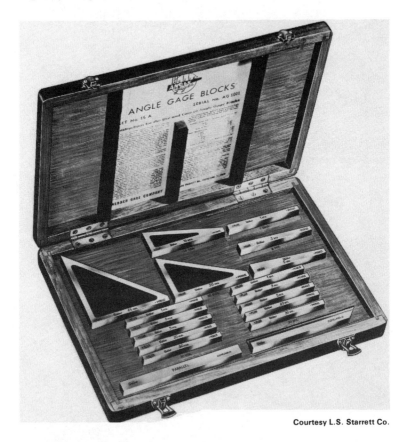

Courtesy L.S. Starrett Co.

Fig. 10-38. A set of angle gage blocks.

Fig. 10-39 shows these blocks being used with a dial indicator to check an angle of 13°. Fig. 10-40 shows these blocks being used to set a revolving magnetic chuck for a 38° angle.

The toolmaker must always keep in mind that improved measuring tools and instruments are being developed, and it is impor-

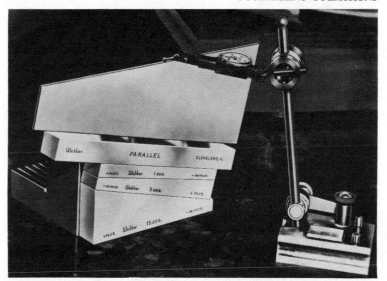

Fig. 10-39. Angle gage blocks checking a 13° angle.

tant to stay abreast of the developments in the area of precise measurement. Measuring tools are available with metric calibrations, and the toolmaker needs to be familiar with this system of measurement. The toolmaker needs to keep informed of all developments in his field, including metric measurement.

SUMMARY

Some of the requirements for becoming an expert toolmaker are the ability to read blueprints, a knowledge of basic mathematics, the ability to read precision measuring tools, and pride in keeping instruments and precision tools clean and in working order.

The art of toolmaking can best be acquired from experience. At first, the apprentice is permitted to perform the more simple operations of toolmaking. As skill and experience are gained, he can be trusted with more intricate work until he can master any work that comes into his department. A good toolmaker is always a neat worker. He also keeps the tools in his department in perfect

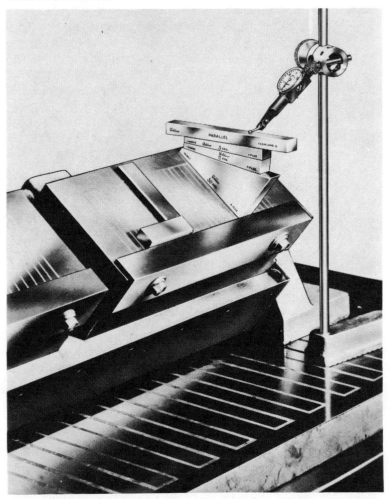

Fig. 10-40. Angle gage blocks being used to set a revolving magnetic chuck.

condition and adjustment because he knows that his workmanship is dependent on their accuracy.

A precision measurement must be accurate to within one-thousandth of an inch or a fraction thereof. These measurements are made with precision tools, which are constructed in such a manner that direct readings can be made of errors in dimension, even though the errors are minute.

Direct contact measurement is important, and the development of "feel" is a must. The toolmaker must understand tolerances and fits as well as proper layout. He must be able to "read the angles" with a variety of tools and select the proper tool for the accuracy needed. Sine bars, sine plates, gage blocks, angle gage blocks, electronic height gages, etc., should be familiar working tools. Also a full working knowledge of the metric system is needed.

REVIEW QUESTIONS

1. What are some of the requirements to become an expert toolmaker?
2. What is a "forced fit" and a "shrink fit"?
3. What are tolerance limits on precision measurements?
4. What is meant by the ABC agreements?
5. What is a sine bar?
6. What are some other methods of measuring angles? How accurate is each?
7. Explain "feel" and its importance to the toolmaker.
8. What is the use of a vernier height gage? A dial indicator height gage? An electronic height gage?
9. How are gage blocks selected for a specific height?
10. How are angle gage blocks selected for a specific angle measurement?
11. How important is the metric system?

CHAPTER 11

Heat-Treating Furnaces

Various types of furnaces have been designed for heat-treating ferrous and nonferrous metals. It is difficult to classify completely the many types of furnaces used in heat-treating processes. The size and shape of the parts, the volume of production to be handled, the type of treatment needed, as well as considerations of economy and efficiency, are all factors to be considered in the selection of heat-treating furnaces.

CLASSIFICATION

In general, heat-treating furnaces can be classified with respect to: (1) source of heat, as gas-fired, oil-fired, and electric; (2) method of applying heat, as overfired, underfired, recuperative, and forced convection; (3) method of heat control, as manual or automatic; (4) protection of the work, as a controlled-atmosphere type of furnace; and (5) batch of continuous operation.

TYPES OF FURNACES

Many types of furnaces are designed for a special application or feature, such as high-speed, pusher, rotary-hearth, tilting-oven, pit, pot, box, tool, air-tempering, continuous high-temperature, and carburizing furnace. Furnaces can also be classified with respect to the maximum temperature for which the furnace was designed as: (1) low (600°F, 315.55°C); (2) medium (1200°F, 648.88°C); (3) high (1800°F, 982.22°C); and (4) extra high (2600°F, 1426.66°C). These temperature ratings are not standard ratings, but they are compiled from a large number of manufacturers' ratings for many types of furnaces. A simple production line might consist of four furnaces, as illustrated in Fig. 11-1. The furnaces in order of use are: preheat, high heat, quench, and draw or tempering. These furnaces would be selected according to the temperature requirements of the work being heat treated. Other installations might include oil or water quench, soak, and/or rinse tanks.

Gas-Fired Oven Furnaces

A sectional view of an oven-type heat-treating furnace is shown

Courtesy A.F. Holden Co.

Fig. 11-1. A four-furnace production setup.

in Fig. 11-2. A section through the burner inlet shows the basic construction of the furnace. Essentially, these furnaces have a substantial frame or casing mounted on cast-iron legs; heavy steel channels are provided to prevent sagging. Many furnaces have a counterbalanced door and lifting mechanism.

Depending on the size of the furnace, the interior walls consist of 4.5- to 5-inch fire brick backed up with insulation. The hearth is made of flat preburned forms of fire clay.

The burners are placed properly to fire underneath the working hearth without impinging on the refractory or workpiece (see Fig. 11-2). This type of burner arrangement is used for heat treatment of the low-carbon steels.

For high-speed temperatures, one set of burners is placed to fire underneath the hearth, and the opposite set is placed to fire directly underneath the roof or arch of the furnace. This burner arrangement provides a rotary action for the hot products of combustion and insures rapid and uniform heating; it also prevents buildup of excessive temperatures beneath the hearth. According to the type of burner equipment provided, the operating temperatures are from 1600°F (871.11°C) to 2400°F (1315.55°C).

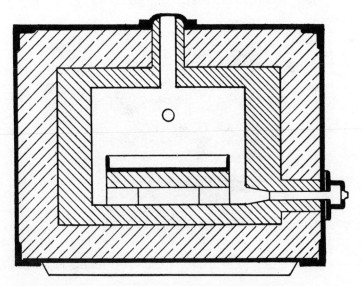

Fig. 11-2. Sectional view of an oven-type heat-treating furnace.

Electrically Heated Furnaces

Heat-treating furnaces can be heated with electricity. It is convenient, does not require a flue, is more accurately controlled, and in certain cases can provide higher temperatures. Electrically heated furnaces can be divided into four classes: (1) open element, where the heating element is exposed to the atmosphere of the furnace; (2) closed element, with the heating element sealed into the walls and/or floor of the heating chamber; (3) immersion type, which has a sealed resistance heating element immersed in a molten bath (usually limited to 1100°F−593.33°C); and (4) electrode type, consisting of electrodes suspended into an electrically conductive salt. The current passes through the salt between the electrodes. This has an added advantage as the electrical currents cause the salt bath to circulate and this tends to maintain a uniform temperature.

Fig. 11-3 shows an open element electric heat-treating furnace with an automatic control pyrometer. Some such furnaces are very large and have a car on which the work is loaded. Then the car is run into the furnace through large doors.

Fig. 11-3. Open element electric heat-treatment furnace.

Fig. 11-4 shows the electrodes in a long furnace without the salt bath. As can be seen, the electrodes are along one side, usually the back side, and extend almost to the bottom. This furnace has four groups with four electrodes each. The number of electrodes depends upon the size of the furnace, the temperature required, and the amount of work to be processed.

Courtesy A.F. Holden Co.

Fig. 11-4. Electrodes of a long salt-bath furnace.

Pit Furnaces

Vertical furnaces of the pit type are used for heating long, slender work. These are sunk into the floor like a hole with tops that can be swung off to open the furnace. Warpage can be minimized by suspending long pieces vertically. Pit-type furnaces

325

can be used to heat batches of small parts, which can be loaded into a basket and lowered into the furnace. Fig. 11-5 shows such a furnace. The parts (bevel gears) are loaded on a rack and lowered into the deep pit furnace. This particular furnace is an atmosphere carburizing type. Pit furnaces can be gas, oil, or electrically heated.

Courtesy Illinois Gear

Fig. 11-5. Deep pit furnace being loaded with bevel gears.

Pot Hardening Furnaces

The pot furnace is designed for indirect heating, the materials being placed in a liquid heat-transmitting medium. This is the immersion method of heat-treating small articles. The immersion method is adapted to lead hardening, liquid carburizing, liquid nitridizing, drawing, and reheating. The pot furnace has become more popular as a tool for industrial production because it is convenient, clean, accurate, economical, speedy, reduces warpage, and eliminates scale.

Pot furnaces are built for either gas, oil, or electrical firing. The basic parts of a pot furnace are: (1) furnace, (2) pot, (3) drain, and (4) hood.

The furnace itself is a circular or rectangular casing with an interior insulated lining. A cast-iron top is usually provided to support the pot. The drain is provided for accidental breakage of the pot and it is sealed by a plug to prevent passage of air into the furnace chamber while in operation. The steel hood is required when the furnace is operated with lead, cyanide, or molten salts, which give off objectionable or dangerous fumes (Fig. 11-6). The hood should be provided with a flue pipe that connects it to the vent. Swinging doors make the pot accessible for charging and discharging.

When the pot furnace is either gas or oil fired, the burners are

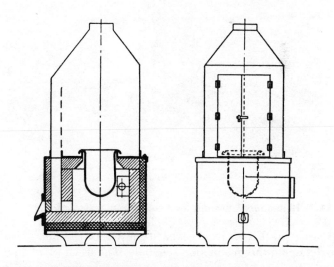

Fig. 11-6. Pot furnace provided with hood for use with lead, cyanide, and other molten-bath heat-treating materials.

327

arranged to fire tangentially through accurately formed combustion tunnels into the combustion chamber (Fig. 11-7). This avoids flame impingement on the pot and provides thorough mixing of combustion gases; it also ensures uniform heating. Burners, vents, and holes for lighting the burners are located to provide for the desired service.

According to the requirements, either manual or automatic control can be provided for the burners. Burner capacity to 1650°F (898.88°C) can be provided for hardening service. For tempering service, temperatures as low as 350°F (176.66°C) can be obtained on the same furnace with low-pressure gas equipment by furnishing two proportional mixers, using only one for low-temperature requirements. Additions are not necessary on high-pressure gas equipment because individual burners can be shut off independently.

Courtesy Johnson Gas Appliance Co.

Fig. 11-7. High-temperature gas-fired pot furnace.

Industry prefers furnaces for each temperature range. This aids in maintaining the production schedule. The usual practice is to provide furnaces for the various temperature ranges required. Thus, a furnace will be used within the temperature range for which it is designed. Fig. 11-8 shows a medium-temperature gas-fired pot furnace. The section drawing shows one method of providing constant circulation within the pot. Other methods include the use of pumps and agitators.

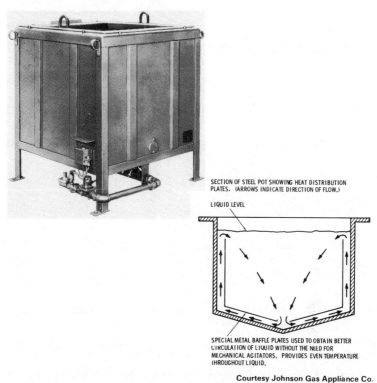

SECTION OF STEEL POT SHOWING HEAT DISTRIBUTION PLATES. (ARROWS INDICATE DIRECTION OF FLOW.)

LIQUID LEVEL

SPECIAL METAL BAFFLE PLATES USED TO OBTAIN BETTER CIRCULATION OF LIQUID WITHOUT THE NEED FOR MECHANICAL AGITATORS. PROVIDES EVEN TEMPERATURE THROUGHOUT LIQUID.

Courtesy Johnson Gas Appliance Co.

Fig. 11-8. Medium-temperature gas-fired pot furnace.

The basic design of an electrically heated pot furnace is shown in Fig. 11-9. A typical heating element is formed of a single continuous helix of extra-heavy nickel chromium rod, either round or square in shape, depending on the size of the furnace. The heating element is precise in shape and spacing and is apportioned around the lining tile of the chamber so that a uniform diffusion of heat is achieved. The size and length of the coil permit a low-watt density and a lower-element temperature for the corresponding temperature of the bath.

329

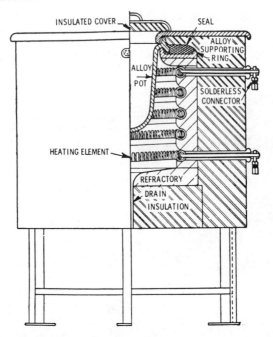

INSULATED COVER   SEAL

ALLOY SUPPORTING RING

ALLOY POT

SOLDERLESS CONNECTOR

HEATING ELEMENT

REFRACTORY

DRAIN

INSULATION

Fig. 11-9. Cutaway view of an electric pot furnace.

The terminals of the heating element are large-diameter rods of the same material brought to the outside of the furnace through special tubes with asbestos packing, which prevent heat loss and short circuits. Solderless bronze connectors are supplied for connection to the power supply.

Many of the pot furnaces in industry are the electrode type. The resistance of the bath to electrical current is used to heat and hold the bath to any desired temperature between 350°F (176.66°C) and 2350°F (1287.77°C). This is an efficient medium-size furnace that is ideal for use with conveying equipment. The control is accurate, and the maintenance cost is low. It is a practical furnace for heat treating high-speed steels. Ceramic pots are used for the higher temperatures. Fig. 11-10 shows two of these furnaces. The one on the left has a low-temperature bath.

Much of the increased use of the electrode-type furnace is due to the development of salt baths to use in the furnaces. Simple tests have been devised to check the baths chemically so the proper rectifiers can be added. In most cases the workpiece can be cleaned by washing in hot water. Control of distortion is feasible and easily accomplished. The salt bath provides support of at least one-fourth of the weight of the part. A salt bath preheat will overcome machine or forming stresses. The salt baths are usually divided into three types: (1) neutral salts, (2) nitrate salts, and (3) cyanide-bearing salts.

A neutral salt bath provides a molten heating medium inert to the workpiece. The neutral heat-treating process is designed to change only the physical characteristics of metal. It is compounded so it will not alter the chemical composition of the work and provides maximum protection for the metal during the heat-treating process.

Courtesy A.F. Holden Co.

Fig. 11-10. Typical installation of two electrode-type pot furnaces.

A nitrate salt bath is used primarily for tempering and operates from 300°F (148.88°C) to 1100°F (593.33°C). Its function is to maintain the metal's physical property or to permit the transformation to a specific microstructure during the cool-down period. It is highly efficient in eliminating distortion and quench cracks in

331

the workpiece. It is useful in processing finished machine parts that must be held to close tolerances.

Cyanide-bearing salt baths are used in heat treating and case-hardening ferrous metals. They can be used to increase the carbon content of low-carbon steels to provide a hardened metal surface to resist wear and abrasion, while the soft interior gives toughness, ductility, and strength to the entire part. The depth of the case can be varied according to the needs of the part. Nitriding, a special process within this category, is producing an extremely hard case on certain types of steel at relatively low temperatures [950°F (509.99°C) to 1200°F (648.88°C).] The process consists of placing the machined, heat-treated, and ground part in a nitrogenous medium for a preselected time at the required temperature. The steel must contain alloying elements such as aluminum, chromium, or molybdenum. The hardness of the case does not depend upon quenching and will retain its hardness at temperatures up to 1200°F (648.88°C). The part can be finished after nitriding by lapping or buffing.

Recuperative Furnaces

A recuperative furnace is designed to recover as much heat as possible from the heated charge while it is cooling. In the common "in-and-out" or batch furnace, a cold charge is placed in the furnace, brought to temperature, and then removed to cool in the air. This method is usually too expensive for modern industry because almost 100 percent of the useful heat is wasted. Such a furnace is especially expensive to operate where slow cooling is necessary because a large quantity of additional heat is required to return the lining of the furnace to operating temperature for each successive charge.

The cost of "in-and-out" or batch-type furnace operations can be reduced slightly by placing an interchanger chamber that is large enough to hold two charges near the batch-type furnace, as shown in Fig. 11-11. A heated charge from the furnace is placed in the interchanger chamber near a fresh untreated charge of material, which receives a partial preheating. Although about one-half the latent heat of the treated charge is available for preheating, the recoverable heat is actually much less because of large losses through the cold furnace walls. A 10 to 15 percent reduction in power consumption can be effected by using the interchanger chamber; but this is not a real saving because the saving is offset by the increased handling cost.

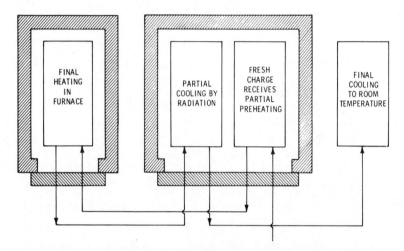

Fig. 11-11. Diagram showing interchanger chamber sometimes used with a batch-type furnace to secure a slight gain in efficiency.

As shown in Fig. 11-12, the interchanger chamber can be built as a single unit for a slightly improved arrangement. This arrangement can result in a 15 to 20 percent saving, which remains as only a slight improvement of the batch-type unit. The recuperative process occurs only in a single stage in this unit. Theoretically, a 50 percent saving is possible with perfect insulation and unlimited time; however, the larger portion of the heat is lost through the walls or retained in the hot charge.

The best results can be obtained by designing the furnace on the counterflow principle—to reduce the loss of heat on the outgoing material. This design is illustrated in Fig. 11-13. As noted in the illustration, a central heating chamber is placed between two recuperative chambers.

The material being heat treated is advanced through these chambers in opposite directions. The ingoing cold material entering the recuperative chamber is preheated by being exposed con-

333

tinually to warmer outgoing material. By the time the ingoing material reaches the heating chamber, it has been heated to a high degree, requiring only a small quantity of heat for a small length of time in the heating chamber to increase the temperature to the desired degree.

As the material moves outward through the second recuperative chamber, it gives up heat that is no longer needed; and by continuous exposure to cooler material, it emerges from the furnace at a lower temperature.

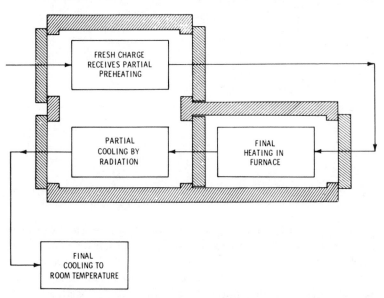

Fig. 11-12. Diagram showing an interchanger chamber and furnaces built as a single unit to improve efficiency.

Fig. 11-13. Diagram of a recuperative type of furnace in which heat is radiated from the finished material to the fresh stock, resulting in maximum heating efficiency.

CONTROLLED ATMOSPHERE

A *neutral atmosphere* is essential for the correct hardening of high-speed tools. Considerable research has been conducted with a wide range of atmospheres to determine the neutral atmosphere and develop methods of achieving it. An understanding of the changes produced by different furnace atmospheres increases the appreciation of the advantages of a truly neutral atmosphere. Three surface changes that can be produced by different furnace atmospheres are: (1) scale, (2) decarburization, and (3) carburization.

Scale

When air enters the heat-treating furnace, the oxygen combines with the steel to produce scale on the surface. Scale always reduces the size of the piece that is being heat treated. Since this burning away of the metal cannot be controlled within close limits, scale is objectionable in the heat treatment of tools that should be held closely to finished size.

The quantity of scale that is formed can be reduced by introducing carbon-bearing compounds or gases. These combine with the oxygen in the air to form carbon dioxide and carbon monoxide gases in proportion to their quantities and rates of combustion. By increasing the richness of the furnace atmosphere, the quantity of scale can be reduced until a point where the scale disappears completely is reached; however, this is not a neutral point because an abrupt change causes the tools to be decarburized rather than being scaled.

Decarburization

This is a most objectionable condition because the loss of carbon causes the surface of the steel to be very soft. A furnace atmosphere that can burn the steel to form scale is also capable of combining with the carbon in the steel to produce decarburization; therefore, it seems unusual that a scale is sometimes formed on high-speed tools, and they are not decarburized. However, the atmosphere can be slightly richer with no scale present, and the tools will become decarburized to a large degree, which is detrimental.

The burning process on the surface of the steel during formation of scale seems to form a glaze that confines the carbon; but if the atmosphere is rich enough to prevent formation of scale, the carbon is free to escape, leaving a soft surface on the steel. This fact seems to be true for almost all high-speed tungsten steels; however, most high-speed steels that contain cobalt or molybdenum can form scale and decarburize at the same time.

After the point where the scale disappears is reached, the richness of the atmosphere can be increased to decrease the amount of decarburization until it disappears, and the neutral point, where the steel is neither giving up nor absorbing carbon, is also reached. The richness of the atmosphere can be increased further to carburize the steel and to increase the rate of carburization, depending on the characteristics of the steel, the furnace temperature, and the gases present.

Carburization

This condition is usually the least objectionable of the three conditions because a slight increase in surface hardness caused by the additional carbon is usually advantageous. The correct furnace atmosphere for hardening high-speed tungsten steel occurs at a point between decarburization and carburization—this is the only satisfactory furnace atmosphere for hardening high-speed cobalt and molybdenum steels. The time element is not critical; tools can be given an ample "soaking" period to assure maximum hardness.

Tools can be heated in a controlled atmosphere for more than an hour at 2350°F (1287.77°C) without any change in size or condition of the surface to indicate the application of excessive heat. The neutral furnace atmosphere envelopes or surrounds the piece that is being heat-treated to protect it.

A precisely controlled atmosphere, maintained in sufficient volume to fill the furnace chamber completely, protects both the surface and the structure of the steel, preventing scale, decarburizing, pitting, and other harmful effects. This, in turn, reduces grinding and finishing costs, improves structure and production life in service, and eliminates spoilage caused by various forms of harmful effects, thereby accomplishing overall savings in tool cost.

CONTROLLED-ATMOSPHERE FURNACES

Various types of controlled-atmosphere furnaces are designed to fullfill the special heat-treating requirements in shops where tools and dies are heat treated in large volumes. These furnaces are also adaptable to shops in which the volume is not large but the tools and dies are of a design that substantial loss could result from inadequate furnace equipment. A diagram of a controlled-atmosphere furnace is shown in Fig. 11-14.

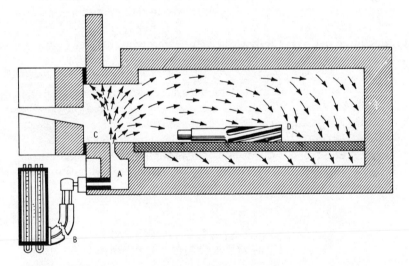

Fig. 11-14. Diagram of a controlled-atmosphere furnace.

An air-gas mixture is precombusted in the precombustion chamber (A in Fig. 11-14) to form a stable protective atmosphere of any desired proportion to fulfill the requirement of a particular type of steel that is being heat treated. A manometer B regulates the air-gas mixture entering the precombustion chamber. The precombusted atmospheric mixture is forced upward through an unbroken slot C across the furnace opening with sufficient pres-

337

sure to seal the opening positively against the entrance of outside air and with sufficient volume to fill the heating chamber— surrounding the workpiece D at all times with a protective atmosphere suitable for the makeup or composition of the steel in the workpiece that is being heat treated.

This method introduces the principle of precombustion of a gas-air mixture and the use of the resulting products of combustion to provide a protective atmospheric medium in the heat-treating furnace. To obtain the controlled atmosphere: (1) the door of the furnace is sealed against the entrance of the outside atmosphere; and (2) an enveloping or furnace atmosphere of any analysis that is desired is provided for protection of the work undergoing heat treatment.

The controlled atmosphere can be varied by turning the valves provided for that purpose to give the desired manometer settings. Once the most effective atmosphere has been determined, it can be duplicated accurately in the future by repeating the settings on the manometer.

A controlled atmosphere was needed in electric furnaces where gas was not used. This led to the use of several inert gases to provide the controlled atmosphere, which has been so successful that they are being used in gas-heated furnaces. It is rather easy if the heating chamber for the work is isolated from the burners. The savings on machining costs after heat treating make it economical to operate. Fig. 11-15 shows atmosphere-controlled carburizing furnaces in use.

TEMPERATURE CONTROL
OF HEAT-TREATING FURNACES

Regulation of the temperature of the workpiece being treated is the basic factor in modern industrial furnace control. The controls must regulate automatically. This must be done at the lowest possible cost in spite of changes in production, ambient temperature, supply voltage, fuel characteristics, or other changes in the heat-treating process.

It is the temperature of the workpiece that is important—not the furnace temperature. The furnace is a basic component of the temperature-regulating system. The temperature of the work is controlled, either directly or indirectly, by regulating the furnace temperature.

r ig. 11-15. Atmosphere-controlled curburized furnaces.

In addition to the furnace, or heating equipment, temperature-control systems include the following parts (Fig. 11-16):

1. The primary element or sensing device.
2. The instrument or controller.
3. The final control element, or power-control device.

Most temperature controls for heat-treating furnaces are referred to (incorrectly) as pyrometers. A pyrometer is a device for measuring the degrees of heat higher than those recorded by a mercurial thermometer. It includes the sensing device and a meter to indicate the number of degrees.

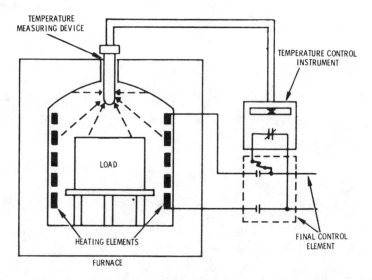

Fig. 11-16. The heating elements of a heat-treating furnace are controlled by relays and switches. Temperature sensing devices cause the relays to energize and thereby turn the power on or off. The types of heat that affect the temperature control are indicated by the arrows.

Results Are Important

The results of the temperature control, rather than the absolute temperature, are more important in measuring a controlled temperature. Correct operating temperature is determined by trial and error and by the results of heat treating—regardless of the temperature indicated by the measuring device.

Absolute temperature is difficult to determine because:

1. Temperature is not constant in most furnaces.
2. The temperature in the various parts of the furnace or work-piece is always different because of differences in losses and distribution of heat throughout the furnace.
3. The temperature-measuring device indicates the furnace temperature at a single point in the furnace rather than the temperature of the workpiece.

Response

The temperature-measuring device continually absorbs or loses heat because furnace temperature changes continually (Fig. 11-17). Accuracy of the device in following the changes depends on its speed in responding. In general, the temperature-measuring device should respond a little more rapidly than the workpiece that is being heat treated.

A measuring device with quick response should be used when thin steel strip is heated; a more sluggish device can be used when heating heavy castings or tightly wound coils. Measuring devices respond more rapidly at higher temperatures.

Measuring Temperature

The heat energy in a body is indicated by temperature.

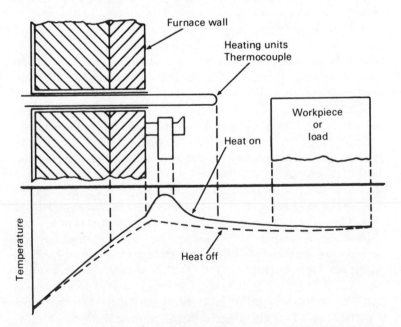

Fig. 11-17. Variations in temperature inside a furnace during the on and off cycles. Note the temperature is greatest next to the heat source; then it tapers off, there being less variation near the workpiece (load) than through the brickwork or insulation.

Temperature rarely can be measured directly, but it is inferred by observation of its effect on a known substance.

All materials are affected by temperature, and an almost unlimited number of methods are used to measure it. The most familiar method is the bimetal thermostat commonly used in home furnace control. It is unsatisfactory for remote indication, and it has limited maximum temperature [around 1100°F (593.33°C)]. A nonindicating unit [up to 1800°F (982.22°C)] can be used for overtemperature protection.

Bulb-type devices—This type of detector uses a metal bulb filled with liquid, liquid and vapor, or gas connected to the operating mechanism through a thin capillary tube. The expansion of the material, transmitted through the capillary tube, operates a bellows, diaphragm, or helix, which, in turn, operates the control mechanism and sometimes an indicating or recording device. The bulbs can be operated at a range of −300°F (−184°C) to +1200°F (648.88°C), depending on the material used in the bulb.

Bulb-type devices are inexpensive, accurate within reason, and commonly used for control applications such as the cooling of water. Higher temperatures are limited chiefly because of the capillary tubing; therefore, the bulbs are not used to great extent for furnace temperature control.

Temperature-sensitive devices—Temperature-sensitive resistors are becoming more common for the higher-temperature ranges. They consist of a small coil of copper, nickel, or platinum wire wound on an insulating form and enclosed by a protective tube. The change in resistance of these materials is uniform and can be predicted for changes in temperature. These devices are among the most accurate that are available if they are used properly. They are particularly adapted to low temperatures or to small differences in temperature. They are very accurate for temperatures below 1600°F (817.11°C), but should not be used continuously for higher temperatures. The maximum temperature is about 3200°F (1760°C). The chief disadvantage of temperature-sensitive resistors in industrial furnace work is the fact that most materials are adversely affected by the common reducing atmospheres. These are usually referred to as resistance pyrometers.

Radiation detectors—The basic components of radiation detectors are the small thermocouples connected in series with their hot junctions close together. A mirror or lens is used to focus radiant energy on the hot junctions. Radiation detectors can be used at temperatures above that usually permitted for standard thermocouples or other devices used to measure the temperature of moving objects, such as steel strips.

The radiation detector's usual temperature range is 800°F (426.66°C) to 3300°F (1815.55°C), but it can be used at temperatures as low as 125°F (51.66°C) and as high as 6000°F (3315.55°C). Low temperatures require a special control for the reference junction. They are difficult to calibrate in terms of absolute temperature.

The practical temperature span and usable range is limited because radiation detectors are nonlinear. They should be used only in the upper half of their calibrated range.

Metal or ceramic tubes of either the open-end or closed-end type are used to protect radiation detectors. The instruments are usually furnished with calibrating rheostats because open-end tubes may have to be calibrated for emittance from the hot body.

Temperature measurements should not be taken at face value. *Radiant energy* that is given off by a hot surface is a function of the surface temperature and surface emittance. These characteristics vary for different materials and even for the same material at different temperatures.

Several problems are involved in measuring the load temperature in a closed furnace. If the load temperature is constant, the radiant energy absorbed must equal that which it emits. If the load is opaque, all the radiant energy falling on it must be absorbed and subsequently emitted, or reflected.

Radiation detectors cannot differentiate between radiated heat and reflected heat (Fig. 11-18). If the temperature of the workpiece is at the same temperature as the furnace walls, it will show black body characteristics, and the temperature reading will be accurate. For the same reason, an accurate furnace wall temperature can be obtained if the volume of work is small compared with the volume of the furnace.

If the temperature of the load is lower than the temperature of the furnace walls, the radiant heat reflected by the work represents a hotter temperature than that of the work, and the temperature reading will be higher than that of the work. Similarly, if the

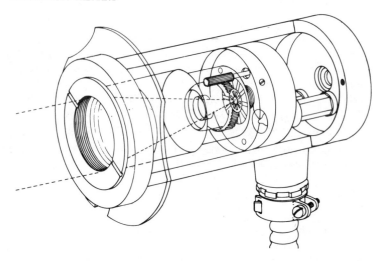

Fig. 11-18. A radiation detector. This type is recommended for furnaces operating at 2700°F (1482.22°C) or above. The radiation detector cannot differentiate between radiant heat and reflected heat.

hot workpiece is surrounded by a temperature lower than its own temperature, the temperature measured will be lower. If the temperature of the furnace wall and emittance of the hot object are known, the error can be calculated. This is difficult to calculate with a high degree of accuracy.

The radiation detector is practically the only satisfactory commercial device for furnaces operated at temperatures above 2700°F (1482.22°C). It is most satisfactory for moving objects or objects that cannot be touched. It is sometimes used to measure the temperature of workpieces heated by induction. For this application, radiation detectors that require a small viewing area and respond quickly (98 percent in 1 second) are available.

Consideration must be given to the effect of discontinuity in measuring and controlling the temperature of moving objects that do not move continuously. The measuring device must be made slow (damped) purposely to prevent a dip in temperature indication each time there is discontinuity. This dip in temperature can cause undesirable control action.

In addition to interpretation difficulties, radiation detectors have other disadvantages in that they are more expensive than most other temperature-measuring devices. The devices are also subject to calibration errors because of errors caused by fogging, changes of emittance, and the effects of gases, smoke, and flames.

These radiation pyrometers are not connected to the furnace. No part of the instrument is inserted in the high heat to be measured. The tube is pointed toward the open door of the furnace or at an opening, and the temperature can be read on the indicator. They can be hand held or mounted on a tripod.

Optical pyrometers—There are several types of optical pyrometers. In one type an electric lamp filament is heated to the same color as that of the part being heated, the temperature of which is needed. The current consumed is indicated, and the corresponding temperature is determined. There are several other types of optical pyrometers. These pyrometers are used to estimate the highest temperatures and may be used for temperatures above 3000°F (1648.88°C).

Thermocouples

The thermocouple is simple, low in cost, and adaptable. It is the most common and most versatile temperature measuring and controlling device. The thermocouple is part of the most commonly used pyrometer known as the thermoelectric pyrometer. The thermocouple is placed in the furnace and connected by wires to a meter, which may or may not be close to the furnace. Proper selection of thermocouples and their uses is complicated; a vast amount of literature has been written on these subjects.

Basic principle—Basically, a thermocouple consists of two wires joined together at one end (hot junction) and connected at the other end to an electric measuring device. When the junctions of the metals are maintained at different temperatures, electricity is produced. A voltage or electromotive force (emf) is produced if the circuit is open, and a current flows if the circuit is closed.

If one of the junctions is kept at a fixed temperature, the emf will give, for any two given metals, a fixed relation to the temperature at the other junction. This combination is known as a thermocouple, or a "couple," and can be used as a temperature measuring and controlling device.

345

The emf and the current produced by the thermocouples are an approximate linear function of the temperature difference between the hot and the cold junctions. Calibration tables are based on a cold junction temperature of 32°F (0°C). The cold junctions are often placed in a thermos bottle of ice water. Most temperature units use room temperature for the cold junction; a compensating resistor or other means is provided for calibration.

The two wires are insulated by porcelain beads in a typical thermocouple (Fig. 11-19). A junction block is provided for the thermocouple and the extension lead wire that connects it to the temperature-measuring instrument. The junction block and the thermocouple are enclosed and held by a head, usually a diecasting.

Protection of thermocouple—Thermocouples are usually enclosed in the modern furnaces to protect them from gases, liquids, and mechanical damage, although open thermocouples are sometimes used. The enclosed thermocouples are supported, and installation in pressure or liquid vessels is simplified.

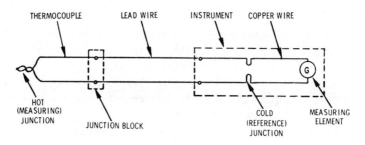

Fig. 11-19. Diagram of a typical electrical circuit for a thermocouple.

The location of the thermocouple is important. Several points should be considered:

1. The thermocouple should be affected only by the heating unit that it controls. Except for protection against over-temperature, it should never be placed between two furnace zones.

2. The thermocouple should be inserted in the furnace far enough to prevent cooling by conduction, but not far enough for the thermocouple to droop. Its calibration should not be

affected by a temperature gradient throughout its length in the furnace.

Thermocouples mounted horizontally in well-insulated furnaces are inserted beyond the inside brickwork a distance of four to six times their outside diameter. Immersion up to ten times the outside diameter may be required for some applications; the depth of immersion should not be changed once the thermocouple is installed.

3. The thermocouple should not be placed near a heating element. Similarly, the thermocouple should be protected from direct contact with the flames in open-fired gas or oil furnaces.

4. When thermocouples are placed in protective atmospheres, the products of combustion should not be allowed to escape past the connection block. Escaping gases can heat up the connections to the lead wire, causing errors in temperature readings; but they can also fill the conduits with explosive gases. The head should be screwed to a pipe nipple that is welded to the furnace casing or otherwise sealed against the loss of furnace atmosphere.

5. The lead wires are color coded, and they should be connected to the correct poles. If the instrument reads in the wrong direction, the connections should be reversed at the connection block rather than at the instrument.

Heat radiation—Virtually all the heat received by the thermocouple is by radiation, at temperatures above 1000°F (537.77°C). In this range, the thermocouple is a radiation detector. Many of the aforementioned observations on radiation apply to detectors.

Calibration—When great accuracy is required, thermocouples can be sent to the National Bureau of Standards in Washington, D.C., for calibration. Most thermocouples are guaranteed for ¾ of 1 percent of the specified calibration at furnace temperatures above 500°F (268.89°C). In some instances, thermocouples made of special-grade wire provide about one-half the standard limit of error.

Many combinations of materials have been developed for thermocouples. Among these materials are: copper-constantan; chromel-constantan; steel-alumel; nickel-tungsten; and others.

The majority of thermocouples used in industrial furnace work are shown in Table 11-1.

Lead wire—Expensive lead wire connecting the thermocouple to the temperature instrument is not necessary. Extension lead wires of the same or more inexpensive materials can be used; these are usually two insulated 16-gage wires in a common outer braid. If it is possible, there should be no joints between the thermocouple and the instrument.

Table 11-1. Basic Thermocouple Types

Type	Temperature Range (°F)	Uses
Iron-constantan	30-1400	Low temperatures; air drawing; copper or brass annealing; ovens
Chromel-alumel	0-2100	Copper annealing; wire enameling; air atmosphere; medium-temperature heat-treating
Nickel-nickel (18% molybdenum)	to 2150	Copper brazing; heat-treating steel; and stainless
Platinum-Platinum-rhodium	to 2800	Forging furnaces; high-temperature heat-treating; laboratory work

If a joint is necessary, the wires should be scraped clean, spliced, and soldered or brazed. The extension wires should be run in grounded conduits for protection; they should not, of course, be run in the same conduit with any other wire or closer than 12 inches to alternating current.

When potentiometer-type instruments are used, thermocouples can be located several hundred feet from the temperature instrument. Millivoltmeter distances cannot be as great. Lead wire of a heavier gage should be used for long distances or two or more lead wires can be connected in parallel.

Testing—The temperature-control instrument can protect furnaces against open thermocouples (or broken thermocouples). Thermocouples should be either changed at intervals determined by experience or tested to protect them against calibration errors.

In testing the thermocouple, the check thermocouple should be inserted in the same tube, or well, or in an adjacent test hole. The check thermocouple should not be used for any other purpose and should be tested frequently against a master standard in a salt pot or small tube furnace. The check thermocouple should be of exactly the same type and size as the one being tested.

Automatic Controls

Pyrometers can be arranged so the moving element of the instrument not only indicates the temperature by its position relative to a scale, but also controls the temperature by regulating the heat supply. Fig. 11-20 shows such a control for an electric furnace that uses solid-state devices and is relatively trouble-free. In some cases even the switching mechanism is solid-state devices. The control can be set for any temperature desired within certain maximum and minimum limits and may be applied to furnaces heated either by gas, oil, or electricity. Fig. 11-21A shows such a unit for a gas-heated furnace, while Fig. 11-21B shows the thermostat control for a low-temperature gas furnace where the temperature variation can be larger.

Courtesy Thermolyn Corp.

Fig. 11-20. Automatic control for an electric furnace.

349

(A) Automatic control.

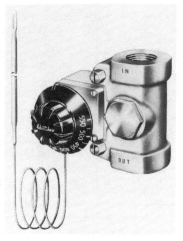

(B) Thermostat control.

Courtesy Johnson Gas Appliance Co.

Fig. 11-21. Gas furnace temperature controls.

Recording Pyrometers

A recording pyrometer is provided with some kind of marking device that traces a line upon a chart graduated with reference to time and temperature. Thus, the temperature at any period is shown graphically on the chart. Such a recording pyrometer may be connected to more than one furnace. Different colors can be used to indicate the different furnaces to avoid confusion.

SUMMARY

In general, heat-treating furnaces can be classified with respect to the source of heat as gas-fired, oiled-fired, and electric; and to the method of applying heat as overfired, underfired, recuperative, and forced convection. The method of controlling heat as manual or automatic and the protection of the work, such as a controlled-atmosphere type of furnace, are also important.

Many types of furnaces are designed for a special application or feature, such as high-speed, pusher, rotary-hearth, tilting-oven, pot, box, tool, air-tempering, continuous high-temperature, and carburizing furnaces. Gas-fired furnaces have a substantial frame or casing mounted on cast-iron legs; heavy steel channels are provided to prevent sagging. Many furnaces have a counterbalanced door and lifting mechanism.

The pot hardening furnace is designed for indirect heating, the materials being placed in a liquid heat-transmitting medium. Pot furnaces are built for either gas or oil firing. The basic parts of a pot furnace are the furnace, pot, drain, and hood. The burners are arranged to fire tangentially through accurately formed combustion tunnels into the combustion chamber.

The basic design of an electrically heated pot furnace is to form a single continuous helix of heavy-duty nickle-chromium rod, either round or square in shape (depending on the size of the furnace), around the lining tile of the chamber so that a uniform diffusion of heat is achieved.

Recent developments have been the improvement of the electrode-type pot furnace and the salt baths used in them. Industry is using more of these furnaces because of their many advantages and low maintenance costs. They are especially useful in the heat treatment of high-speed steel.

A neutral atmosphere is essential for the correct hardening of high-speed steel tools. Three surface changes that can be produced by different furnace atmospheres are scale, decarburization, and carburization. Considerable research has been conducted with a wide range of atmospheres to determine the neutral atmosphere and develop methods of achieving it. Inert gases are being used, especially in electrically heated furnaces, to exclude air from the work being heated.

Much improvement has been made in the controls of heat-treating furnaces. Most such furnaces in industry are being controlled by automatic controls with pyrometers. The desired temperature is set on the control, and the pyrometer holds the furnace at that temperature. A recording pyrometer can also be used if a record is needed of the temperature achieved.

REVIEW QUESTIONS

1. Classify heat-treating furnaces as to source of heat, method of applying heat, method of controlling heat, and protection of work.

2. Explain the basic operations of each type of heat treating furnace.
3. What is controlled atmosphere?
4. What is the advantage of an inert-gas furnace atmosphere?
5. Explain the action of a thermocouple.
6. Explain the action of an automatic control with a pyrometer.
7. What are the advantages of the electrode-type furnace?
8. What types of salt baths are in use?
9. Explain the nitriding process.

CHAPTER 12

Annealing, Hardening, and Tempering

In the heat-treatment process of steel, certain changes occur to alter some of its properties. The processes involve heating and cooling the metal, in its solid state, for the purpose of changing its mechanical properties. Steel may be made harder, stronger, tougher, or even softer through various heat-treating processes. The type of steel and its characteristics must be known before it is heat treated. The material used to make a certain part was selected because of the properties it would possess when the part was finished. (The reader is advised to review Chapter 7 of Volume 1 of Audel's Machinists Library, entitled *Basic Machine Shop*.)

The correct temperature is the important factor in any heat-treating process; improper temperatures in heating do not produce the desired results. Steel will be "burned" at a degree of temperature that is too high. The heat-treating processes are : (1) annealing, (2) hardening, and (3) tempering.

The *degree of hardness* of steel can be changed by heat treatment. Hardness is a relative term. All steels have hardness, but some steels are harder than others.

The heat-treating processes consist of heating metal according to a time-temperature cycle, which consists of three steps:

1. Heating the steel to a certain temperature.
2. Holding the steel at a certain temperature for a period of time for soaking.
3. Cooling at a certain rate.

During these three steps, the properties of the steel may be altered in various ways, depending upon its chemical content and upon the process used. These operations usually alter the internal structure in some way.

ANNEALING

Annealing is a process of softening metal by heating it to a high temperature and then cooling it slowly. Steel is annealed by heating to a low-red heat and letting it cool slowly. The objective of the annealing process is to remove stresses and strains set up in the metal by rolling or hammering, so that it will be soft enough for machining. The longer the period that the metal is allowed to cool, the softer the metal will be because more of the stresses and strains will have been removed.

Methods of Annealing

A common method of annealing steel is to place the steel in a cast-iron box, covering it with a material, such as sand, fire clay, ashes, etc., and then heating in a furnace to proper temperature. The box and contents are then cooled slowly to prevent any hardening. The sand, clay, etc., are placed around the steel to exclude air and to prevent oxidation as well as to delay the cooling process.

Annealing with water is not the best of the annealing processes, but it is a quick method. In this process, the metal is heated slowly to a cherry red. Then it is placed in a dark place and observed until the color is no longer visible; when it reaches this stage, it is cooled in water. The annealed metal is then soft enough to machine. Usually a piece of steel annealed in this way is much softer than if it is packed in charcoal and cooled overnight.

In machining steel that has been annealed by the common method—that is, charcoal—the chips generally are removed in long close-curled lengths, and the surface presents a torn texture. The chips are torn because the steel is too soft; water annealing overcomes this defect. The water annealing method is not generally recommended because it is considered to have a deteriorating effect on the steel.

Another method quite often used is to heat the steel to the proper temperature, holding it at this temperature for the right length of time and then allowing it to cool slowly (often in the furnace) until it is under 1000° F (538°C). Then it is taken out and allowed to cool in air. In general, the higher the carbon content, the slower the cooling rate required.

Temperature for Annealing

The temperature required for annealing of steel is slightly above the *critical point* (the temperature at which internal changes take place in the structure of the metal) of the steel. The critical point varies, of course, with the different steels. Some typical annealing temperatures are: low-carbon steel, 1650°F (899°C); high-carbon steel, 1400°F (760°C) to 1500°F (815.55°C); and high-speed steel, 1400°F (760°C).

Critical points—There are two critical points in steels of ordinary carbon content. A lower temperature is required to produce the internal changes in structure in a steel containing a higher percentage of carbon than in a steel that contains a lower percentage of carbon. The two critical points are: (1) *decalescence* and (2) *recalescence*.

The *point of decalescence* occurs at the temperature where the pearlite changes to austenite as the steel is heated. The *point of recalescence* occurs as the steel is cooled slowly—the austenite returns to pearlite. Thus, the point of decalescence occurs as the temperature is rising, and the point of recalescence occurs as the temperature is falling.

355

Structure of carbon steel—In fully annealed carbon steel, there are two constituents: the element iron in a form known as *ferrite* and the chemical compound iron carbide, known as *cementite*, which has 6.67 percent carbon. Some of the two constituents will be present in a mechanical mixture known as *pearlite*. This mechanical mixture consists of alternate layers of ferrite and cementite. It often has the appearance of mother-of-pearl under the microscope, hence its name. Pearlite contains 0.85 percent carbon, so the amount of pearlite depends upon the carbon content of the steel. A low- or medium-carbon steel will consist of pearlite and ferrite. A carbon steel of 0.85 percent carbon is all pearlite. A high-carbon steel over 0.85 percent carbon will consist of pearlite and cementite.

To fully anneal carbon steel, it is heated above the lower critical point (point of decalescence). The alternate bands of ferrite and cementite (which make up the pearlite) begin to merge into each other. This heating process continues until the pearlite is "dissolved," forming what is known as *austenite*. If the temperature of the steel continues to rise, the excess ferrite or cementite will begin to dissolve into the austenite until only austenite will be present. This temperature is called the upper critical temperature.

If this steel (all austenite) is now allowed to cool slowly, the process of transformation that took place during the heating will be reversed, but the upper and lower critical temperatures will occur at somewhat lower temperatures. The steel at room temperatures will have the same proportions of ferrite or cementite and pearlite as before. The austenite will have disappeared. Fig. 12-1 shows the critical temperatures and the effect of heat upon carbon steel.

Steels with less than 0.85 percent carbon need to be heated above the upper critical temperature in order to be annealed or hardened. This is evident when one locates the "full annealing and hardening range" (in Fig. 12-1) and compares it to the upper critical temperature.

Effects of Forging

Forging is the process by which steel (or other metal) is *heated and hammered* into the various shapes of tools, machine parts, etc.

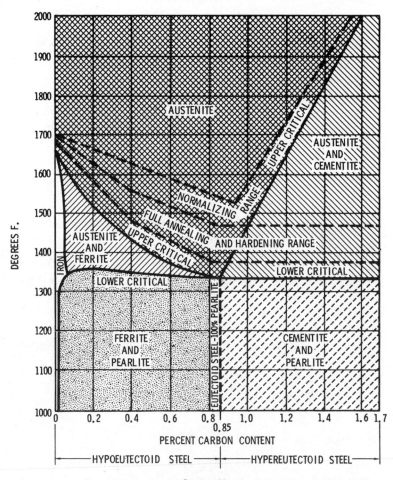

Courtesy Machinery's Handbook, the Industrial Press

Fig. 12-1. Structural constitution of carbon steel.

In the forging process, the piece that is to be forged should be heated uniformly at the correct temperature. The correct degree of heat depends on the kind of steel and the shape, size, and use that is to be made of the finished product.

The hammering process increases the density of the grains in the

357

steel, which improves the steel. Forging should not take place below a certain temperature because the grains can be crushed, which is damaging to the steel.

If the entire piece of work is not forged, internal strains and stresses are produced in the metal because of the difference in densities of the hammered and unhammered portions of the steel. These strains can be relieved by annealing. If only a portion of a piece is to be forged, the piece should be annealed prior to harden- ing and tempering. Examples of pieces that are only partially forged are lathe tools, cold chisels, and screwdrivers; these pieces require annealing, hardening, and tempering.

HARDENING

The process of *hardening* metal is accomplished by increasing the temperature of a metal to its point of decalescence and then quenching it in a suitable cooling medium. In actual practice, the temperature of the metal that is to be hardened should be increased to slightly above the point of decalescence for two reasons: (1) to be certain that the temperature of the metal is above the point of decalescence at all times; and (2) to allow for a slight loss of heat while transferring the metal from the furnace to the quenching bath. Steel with less than 0.85 percent carbon needs to be heated above its upper critical temperature.

Heating Process

Steel can be heated most successfully by placing it in a cold furnace and then bringing the furnace and its charge to the harden- ing temperature slowly and uniformly. Commercially, only those steels that are difficult to harden are heat treated in the manner described above. For example, some tool steels are heated for hardening by first placing them in a preheating furnace at a temperature of 1000°F (537.77°C) or slightly lower. After the steel is heated uniformly to furnace temperature, it is withdrawn and placed in the high-heat furnace at the hardening temperature. When the steel has become heated uniformly at the hardening temperature, it is then quenched to harden it.

During the heating process, the metal absorbs heat, and its temperature rises until the point of decalescence is reached. At this point, additional heat is taken up by the metal. This heat is converted into work to change the pearlite into austenite without an increase in temperature until the process is completed.

This phenomenon can be compared to the *latent heat of steam* in that when the temperature of water is increased to the boiling point, it will absorb an additional quantity of heat without an increase in temperature. The heat is converted into work that is necessary: (1) to bring about a change of state from a liquid to a gas, and (2) to overcome the pressure of the atmosphere in making room for the steam. The total latent heat required to bring about these changes consists of *internal* latent heat and *external* latent heat.

After the metal has been heated, *quenching* serves to fix permanently the structural change in the metal that causes the metal to remain hard after it has been heated to the point of decalescence. If the metal were not quenched and allowed to cool slowly, the austenite would be reconverted to pearlite as the temperature decreases, causing the metal to lose its hardness. When steel is cooled faster than its critical cooling rate, which is the purpose of quenching, a new structure is formed. The austenite is transformed into *martensite*, which has an angular, needlelike structure and a very high hardness.

Martensite has a lower density than austenite; therefore, the steel will increase in volume when quenched. Some of the austenite will not be transformed during quenching. This austenite will gradually change into martensite over a period of time. This change is known as "ageing." This "ageing" results in an increased volume, which is objectionable in many items, such as gages. The cold-treating process (covered in Chapter 16 of this book) can be used to eliminate this problem.

The experienced worker can determine the critical point (point of decalescence) on the basis of color. The heated metal should be transferred to a dark place to judge color. Also, metal bars will decrease in length when quenched at a temperature below the critical point in water. If a metal bar is quenched in water at a temperature above the critical point, hardening also will be indicated.

Pyrometers are used in production work to indicate the critical temperature if the critical temperature of a given steel is known. A magnetic needle can be used to determine whether steel has been heated above the critical point. When heated above the critical point, a piece of steel loses its magnetism; it will attract a magnetic needle if it has been heated to any temperature below the critical point. In making the test with a magnetic needle, the magnetic influence of the cold tongs should be eliminated. A pivoted bar magnet can be introduced into the furnace momentarily to determine the presence or absence of magnetism in the piece of steel.

Heating Baths

Other methods of heating are used in the hardening process. Various liquid baths are used as follows: (1) lead, (2) cyanide of potassium, (3) barium chloride, and (4) mixture of barium chloride and potassium chloride or other metallic salts.

The chief advantage of the liquid bath is that it helps to prevent overheating of the workpiece. The work cannot be heated to a temperature higher than the temperature of the bath in which it is heated. Other advantages include: (1) the temperature can be easily maintained at the desired degree; (2) the submerged steel can be heated uniformly; and (3) the finished surfaces are protected against oxidation.

The *lead* bath cannot be used for high-speed steels because it begins to vaporize at 1190°F (643.33°C). The *cyanide of potassium* bath is used in gun shops extensively to secure ornamental effects and to harden certain parts.

A thermoelectric pyrometer can be used to indicate the temperature of a *barium chloride* bath. *Potassium chloride* can be added for heating to various temperatures as follows: (1) for temperatures from 1400°F (760°C) to 1650°F (899°C), use three parts barium chloride to one part potassium chloride; (2) for higher temperatures, reduce the proportion of potassium chloride.

Quenching or Cooling baths

Although water is one of the poorest of all liquids for conducting heat, it is the most commonly used liquid for quenching metals in heat-treating operations. Water cools by means of evaporation.

To raise the temperature of 1 pound of water from 32°F (0°C) to 212°F (100°C) requires 180 Btu; to convert it into steam, an additional 950 Btu are required. Thus, the latent heat of vaporization (950 Btu) absorbed from the hot metal is the cooling agency. For efficient cooling, the film of steam must be replaced immediately by another layer of water—this requires circulation. If the heated metal is plunged into the water, thermocirculation results; however, if the metal is placed in the water horizontally, the film of steam that forms on the lower side of the piece is "pocketed," and the cooling action is greatly retarded. In still-bath quenching, a slow up-and-down movement of the work is recommended.

In addition to plain water, several other solutions can be used for quenching. Salt water, water and soap, mercury, carbonate of lime, wax, and tallow can also be used. Salt water will produce a harder "scale." The quenching medium should be selected that will cool rapidly at the higher temperatures and more slowly at the lower temperatures. Oil quenches meet this requirement for many types of steel. There are various kinds of oils employed, depending on the nature of the steel being used.

The quenching bath should be kept at a uniform temperature so that successive pieces quenched will be subjected to the same conditions. The next requirement is to keep the bath agitated. The volume of work, the size of the tank, the method of circulation, and even the method of cooling the bath need to be considered in order to produce uniform results.

High-speed steel is usually quenched in oil. However, air is used for quenching many high-speed steels. Air under pressure is applied to the work, that is, by air blast.

TEMPERING

The purpose of *tempering,* or "drawing," is to reduce brittleness and remove internal strains caused by quenching.

The process of tempering metal is accomplished by reheating steel that has been hardened previously, and then quenching to toughen the metal and to make it less brittle. Unfortunately, the tempering process also softens the metal.

Tempering is a reheating process—the term "hardening" is

361

often used erroneously for the tempering process. In the tempering process, the metal is heated to a much lower temperature than is required for the hardening process. Reheating to a temperature between 300°F (149°C) to 750°F (399°C) causes the martensite to change to *troostite*, a softer but tougher structure. A reheating to 750°F (149°C) to 1290°F (699°C) causes a structure known as *sorbite*, which has less strength than troostite but much greater ductility. See Table 12-1 for tempering temperatures for various tools.

Table 12-1. Typical Tempering Temperatures for Certain Tools

Degrees Fahrenheit	Degrees Celsius	Temper Color	Tools
380	193	Very light yellow	Lathe center cutting tool for lathes and shapers
425	218	Light straw	Milling cutters, drill, and reamers
465	241	Dark straw	Taps, threading dies, punches, dies, and hacksaw blades
490	254	Yellowish brown	Hammer faces, shear blades, rivet sets, and wood chisels
525	274	Purple	Center punches and scratch awls
545	285	Violet	Cold chisels, knives, and axes
590	310	Pale blue	Screwdrivers, wrenches, and hammers

Color Indications

Steel that is being heated becomes covered with a thin film of oxidation that grows thicker and changes in color as the temperature rises. This variation in color can be used as an indication of the temperature of the steel and the corresponding temper of the metal.

As the steel is heated, the film of oxides passes from a pale yellow color through brown to blue and purple colors. When the desired color appears, the steel is quenched in cold water or brine. The microscope can be used to explain the phenomena associated with the change in color. Steel consists of various manifestations of the same compound rather than separate compounds.

Although the color scale of temperatures has been used for many years, it gives only rough or approximate indications, which vary for different steels. The color scale for temper colors and corresponding temperatures is given in Table 12-2.

Table 12-2. Temper Colors of Steel

Colors	Temperatures (Fahrenheit)	(Celsius)
Very pale yellow..................	430	221
Light yellow.......................	440	227
Pale straw-yellow	450	232
Straw-yellow	460	238
Deep straw-yellow	470	243
Dark yellow	480	249
Yellow-brown	490	254
Brown-yellow	500	260
Spotted red-brown	510	266
Brown-purple	520	271
Light purple	530	277
Full purple	540	282
Dark purple	550	288
Full blue..........................	560	293
Dark blue	570	299
Very dark blue	600	316

Specially prepared tempering baths equipped with thermometers provide a more accurate method of tempering metals. These are much more accurate than the color scale for tempering.

Casehardening

The casehardening operation is a localized process in which a hard "skin," or surface, is formed on the metal to a depth of $\frac{1}{16}$ (0.0625) to $\frac{3}{8}$ (0.375) inch. This hard surface, or "case," requires two operations: (1) carburizing where the outer surface is impregnated with sufficient carbon, and (2) heat treating the carburized

363

parts to obtain a hard outer case and, at the same time, give the "core" the required physical properties. Casehardening usually refers to both operations.

Carburizing is accomplished by heating the work to a temperature below its melting point in the presence of a material which liberates carbon at the temperature used. The material can be solid (charcoal, coke, etc.), liquid (sodium cyanide, other salt baths), or gas (methane, propane, butane). Often only part of the work is to be casehardened. The four distinct ways by which casehardening can be eliminated from portions of the work are:

1. Copperplating.
2. Covering the portion that is not hardened with fire clay.
3. Using a bushing or·collar to còver the portion that is to remain soft.
4. Packing with sand.

An article that is to be casehardened can be copperplated on the portion that is to remain soft. This is especially useful when a liquid carburizing process is used. The portion to remain soft can also be protected by covering with fire clay, covering with a collar or bushing, or packing the portion in sand. The size and shape of the work will dictate the methods as well as the type of steel and the process to be used.

The casehardening furnace must provide a uniform heat. Steel that is to be casehardened must be selected carefully. Since oil and gas have superseded coal as fuels for casehardening furnaces, furnace construction has changed considerably. Careful consideration should be given the carbonaceous material used in packing the parts and the box in which the material is packed. The operation of packing the insulated workpiece is referred to as "local hardening."

In the casehardened process, articles that are to be casehardened are heated to a cherry red color in a closed vessel along with the carbonaceous material and then quenched suddenly in a cooling bath. Malleable castings can be casehardened so that they acquire a polish. Malleable iron can be casehardened by heating it to a red heat, rubbing cyanide of potassium over the surface or immersing it in melted cyanide, reheating, and then quenching it in water. Usually the hardening operation is a separate operation following the carburizing and is designed for the part and the steel used.

Both iron and steel can be casehardened, but it is used mostly on steel products. The gears of the transmissions of automobiles are a typical example of casehardening, so that they can withstand the abuse of "shifting gears" (not synchromesh) without waiting for them to synchronize.

Variations on Casehardening Methods

A commonly used method of casehardening is to first carburize the material and then allow the boxes to cool with the work in them, after which they are reheated and hardened in water. For work such as bolts, nuts, screws, etc., it is satisfactory to dump them into water directly from the carburizing furnace without reheating.

A common iron wheelbarrow with two pieces of flat iron placed across it lengthwise should be provided. Place a sieve made of ⅛-inch wire, having ¼-inch mesh and approximately 18 inches square by 6 inches in depth, on the bars. The sieve should have a handle ⅝ inch in diameter by 6 feet in length. The boxes are emptied into the sieve, and, after sifting, the heated material is dumped into a tank of cold water, which should be large enough to prevent the water heating too quickly. Care should be taken not to empty the entire contents of the boxes into the water in one place.

. A constant flow of water should be available while the work is being hardened. The work should never be removed from the furnace until the temperature has been lowered. The steel should be treated as tool steel after it is carburized; it is harmful to the steel to harden it at the high carburizing temperature.

Gears and other parts that should be tough, but not extremely hard, should be hardened in an oil bath. The work is less liable to warping, and the hardened product can withstand the shocks and severe stresses without breakage.

SUMMARY

The degree of hardness of steel can be changed by heat treatment. All steels have hardness, but some steels are harder than others. The heat-treating processes of steel are annealing, hardening, and tempering.

Annealing is a process of softening metal by heating it to a high temperature and then cooling it very slowly. The objective of the annealing processes is to remove stresses and strains set up in the metal by rolling or hammering so that it will be soft enough for machining.

The process of hardening is accomplished by increasing the temperature of a metal to the point of decalescence and then quenching it in a suitable cooling medium. The process of tempering is accomplished by reheating steel that has been hardened previously and then quenched to toughen the metal and make it less brittle. Tempering processes also can soften certain metals.

REVIEW QUESTIONS

1. What is the process of annealing, hardening, and tempering?
2. What is the difference between hardening and tempering?
3. What is the point of decalescence?
4. What is the point of recalescence?
5. What happens at the lower critical temperature? The upper critical temperature?
6. What are the following: ferrite, cementite, pearlite, austenite, martensite, trootstite, sorbite?
7. Explain the difference between hardening and tempering?
8. What two operations are involved in casehardening? Which is done first?
9. What mediums are used for quenching?

CHAPTER 13

Principles of Induction Heating

Induction heating is being used extensively in the hardening of steel since it is particularly applicable to parts that require localized hardening or controlled depth of hardening and to irregularly shaped parts that require uniform surface hardening around their contour, such as cams. Some advantages of induction heating and hardening are: (1) little oxidation or decarburization; (2) exact control of depth and area of hardening; (3) good regulation of degree of hardness obtained by automatic timing of heating and quenching cycles; (4) a short heating cycle, which may range from a fraction of a second to several seconds; (5) minimum warpage or distortion; and (6) possibility of using carbon steels instead of more expensive alloy steels.

The designer understands the advantage of applying hardening by induction heating to localized zones. Specific areas of a given part can be heat treated separately. Welded or brazed assemblies can be built up prior to heat treating if only internal surfaces or

367

projections require hardening. Stresses at any given point can be relieved by local heating. It is important for the toolmaker to understand the principles of induction heating.

A magnetic field is set up around a wire that carries an electrical current (Fig. 13-1). If the wire is formed into a coil or a loop, the magnetic field of the coil is intensified (Fig. 13-2).

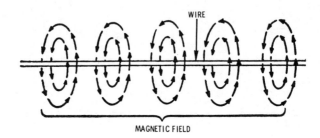

Fig. 13-1. Diagram showing the magnetic field created around a wire-carrying current.

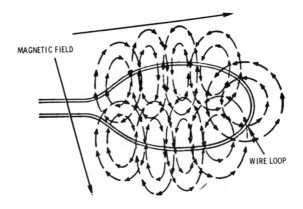

Fig. 13-2. Diagram showing the increase in intensity of the magnetic field when a coil or loop is introduced in the wire carrying a current.

The intensity of the magnetic field is influenced by three factors: (1) the amount of current flowing through the wire; (2) the number of turns or loops in the coil; and (3) the medium within and surrounding the coil (Fig. 13-3).

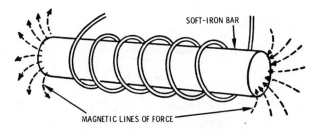

SOFT-IRON BAR

MAGNETIC LINES OF FORCE

Fig. 13-3. Diagram showing the increase in intensity of the magnetic field when the wire carrying the current is wound around a soft-iron rod.

Since the number of turns in the coil influences the intensity of the magnetic field, it is logical that a coil with several turns carrying a small current can produce the same effect as a coil with a single turn carrying a large current. Likewise, if the medium is a poor conductor of magnetic lines of force—air, for example—the magnetic field is weak. The magnetic field is intensified if the medium is a good conductor of magnetic lines of force. When a soft-iron bar is placed in a coil that is carrying a current, an electromagnet is produced (see Fig. 13-3).

An alternating current flowing through the coil of wire causes the magnetic lines of force to expand and contract. If the magnetic lines of force cut an object that is a conductor of electricity, voltage is induced in the object as the lines of force expand and contract. The transformer, which is used for the purpose of either increasing or decreasing voltage, is an example of a practical application of the preceding phenomenon (Fig. 13-4). In the transformer, a secondary coil having a fixed number of turns of wire is placed in the magnetic field of a primary coil having a different number of turns of wire. When a low-frequency alternating current is passed through the primary coil of the transformer, an alternating current of different voltage is induced in the secondary coil. The characteristics of the induced current are determined by the number of turns of wire in the two coils.

Induction heating is a result of application of the aforementioned transformer principle. Since the object to be heated is made of metal and is a conductor of electricity, electrical current can be induced within the object in the same manner that it is induced

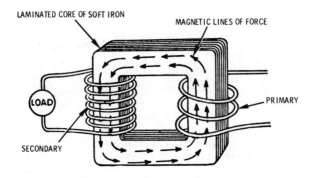

LAMINATED CORE OF SOFT IRON

MAGNETIC LINES OF FORCE

LOAD

PRIMARY

SECONDARY

Fig. 13-4. Note the effect of the laminated soft-iron core on the magnetic lines of force indicated by arrowheads. This is the basic principle of a transformer.

the secondary coil of a transformer. The resistance of the metal to the flow of these induced currents results in rapid heating action that is often called "eddy current losses" (Fig. 13-5). If the object is made of steel or iron, it is magnetized, demagnetized immediately, and remagnetized in the opposite direction, as the current alternates in direction.

At the high frequencies used in induction heating, molecular friction is created by the rearrangement of the steel or iron molecules, as the magnetic state is changed rapidly and continuously. This molecular friction generates heat within the object; the phenomenon is known as heating by *hysteresis* losses. Heat energy can be induced in a piece of steel at the rate of 100 to 250 Btu per square inch per minute by induction heating. This compares with a rate of 3 Btu per square inch per minute for the same material at room temperature when placed in a furnace with a wall temperature of 2000°F (1093.33°C). This means a very short heating cycle is needed.

Usually, the higher the frequency used, the shallower the depth of heat penetration. Low power concentrations and low frequencies are usually employed for heating clear through, for deep hardening, and for large workpieces. Currents at high frequencies are used for shallow and controlled depths of heating, such as surface hardening, and in localized heat treating of small workpieces.

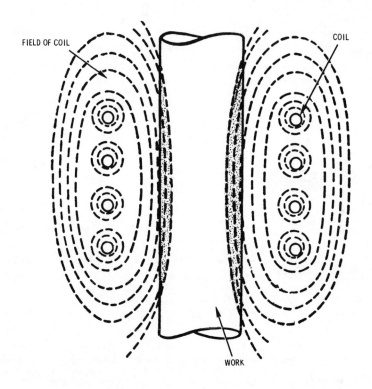

FIELD OF COIL

COIL

WORK

Fig. 13-5. Diagram showing the eletromagnetic field created by a load coil in induction heating.

The depth of heat penetration and the frequency used can be illustrated. A ½-inch round bar of hardenable steel can be heated through its entire structure by an induced current of 2000 hertz so it can be hardened, but a ¼-inch bar of the same steel would not reach a sufficiently high temperature to permit hardening. At 9600 hertz the ½-inch bar can be surface hardened to a depth of 0.100 inch, while the ¼-inch bar can be through hardened. The ¼-inch bar would require a current at over 100,000 hertz for surface hardening.

Hardening, annealing and normalizing by induction, low power concentrations, and low frequencies are desirable. This will prevent too great a temperature difference between the surface and

371

the interior of the work. Widening the spacing between work and applicator coil is one way of reducing the amount of power delivered to the work.

Various shapes of heating coils that are used for high-frequency induction heating are illustrated in Fig. 13-6. Alternating current

Fig. 13-6. Various types of coil shapes used for high-frequency induction heating.

has a tendency to concentrate on the surface of a conductor. This tendency is not noticeable with 50- or 60-hertz current, but it is very pronounced with higher frequencies. Since the center of the conductor carries a very small amount of current, the load coil can be made of hollow copper tubing in which water can be circulated to carry away heat that is generated by the flow of the current and radiant heat from the heated object.

An automatic rod heating device prior to forming is shown in Fig. 13-7.

SUMMARY

Induction heating has come into use because of the need to have local hardening or controlled depth of hardening. There are several advantages of induction heating and hardening.

The intensity of a magnetic field is influenced by three factors; the amount of current flowing, the number of turns of wire in a coil, and the atmosphere within and surrounding the coil.

Since the number of turns in the coil influences the intensity of the magnetic field, it is logical that a coil with several turns carrying a small current can produce the same effect as a coil with a

Fig. 13-7. An automatic rod heating device, using radio frequency induction heating.

single turn carrying a larger current. The magnetic field is intensified if the medium through which it passes is a good conductor of magnetic lines of force. When a soft-iron bar is placed in a coil that is carrying current, an electromagnet is produced.

If the current is alternating, the electromagnet becomes a transformer. Induction heating is an application of the transformer principle. In general, the higher the frequency, the shallower the depth of heat penetration. High-frequency induction heating is a good way to surface harden a piece of hardenable steel.

REVIEW QUESTIONS

1. What three factors influence the intensity of a magnetic field?

373

2. Name three materials that are good conductors.
3. What is a magnetic field?
4. Explain the difference between a primary and a secondary winding.
5. What are the advantages of induction heating and hardening?
6. How is the heating coil kept cool?

CHAPTER 14

High-Frequency Induction Heating

Operations such as annealing, hardening, brazing, soldering, and melting can be performed efficiently with induction-heating equipment. Certain characteristics of alternating current (ac) must be considered for an understanding of high-frequency heating.

The amount of current (amperes) changes constantly as ac flows in a wire or conductor. The amount of current fluctuates continually from zero to a positive maximum and back to zero again; then the amount of current changes to negative maximum and returns to zero. This complete reversal of current is called a *cycle*. If this cycle occurs in 1 second, it is called a *hertz*. The number of cycles that occur in 1 second is referred to as the *frequency* of the current. The frequency of current is usually 60 hertz (cycles per second). High-frequency heating simply indicates that the number of hertz (frequency) of the current supplied normally from the power lines of the utility company has been greatly increased by means of a generator, spark gap, or vacuum-tube equipment (Fig. 14-1).

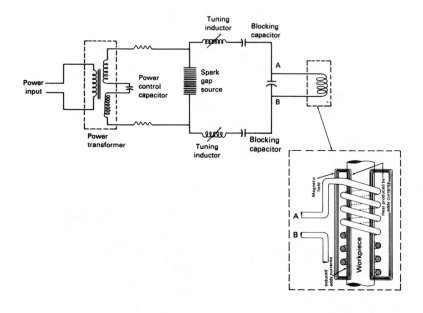

Fig. 14-1. Circuit for high-frequency heat-treating furnace.

PRODUCING HEAT BY RESISTANCE

Almost all objects that are electrical conductors can be heated if they are exposed to a strong electromagnetic field. This field is produced by a strong electric current flowing through the load coil.

In general, metals are good conductors of electricity, and their surface resistance is of such a nature that an appreciably high current (eddy current) can flow at the surface of the workpiece. These eddy currents are dependent on the frequency of the equipment, the surface resistance of the metal, and the power applied. If any of these factors is increased, a greater quantity of heat is produced in the metal. Low frequencies require a close coupling between the coil and the work for transferring energy; high frequencies permit a wider spacing between the coil and the load.

Most of the standard types of steel that can be hardened can be hardened by induction heating. Low-carbon steels with a carburized case, as well as medium- and high-carbon steels, may be used in this process.

HEATING UNITS

High-frequency current advanced from the hypothetical stage to practical realization as early as 1842, when a Professor Henry discovered that the discharge of an electrical condenser across a gap was oscillatory. Years later, as Marconi was making use of this important discovery in developing wireless telegraphy, Egbert von Lepel, a noted scientist, recognizing the shortcomings of spark gaps used by Henry and Marconi, invented what he called the quenched spark gap.

Frequencies of 1000 to 15,000,000 hertz are used in high-frequency induction-heating units. Alternating current, because of its inductive effect, tends to concentrate on the surface of a conductor. This tendency is low with 60-hertz current, but it becomes very pronounced at the higher frequencies of induction heating.

Induction-heating processes are divided into low-frequency and high-frequency types. For several years, 60-hertz current and currents with frequencies up to a few thousand hertz were used in furnaces to melt metal. Newer installations in industry are mostly of the high-frequency type (see Fig. 14-2).

There are three types of equipment used by industry to produce high-frequency current for induction heating: (1) motor generator sets, which provide current at frequencies between 1000 and 10,000 hertz; (2) spark gap oscillator units for frequencies between 80,000 and 300,000 hertz; and (3) vacuum-tube oscillator sets for frequencies ranging from 350,000 to 15, 000,000 hertz or more (see Fig. 14-3).

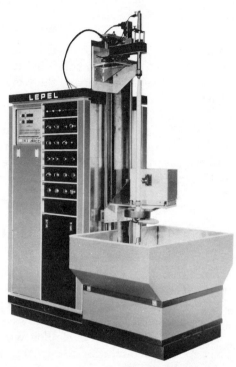

Courtesy Lepel Corporation.

Fig. 14-2. Automatic vertical scanner for hardening rods, shafts, axles, and spindles.

HIGH-FREQUENCY APPLICATIONS

High-frequency induction heating reduces the time required for annealing, hardening, stress relieving, brazing, soldering, and forging of metals from several hours to a few minutes or seconds (see Fig. 14-3). In the process of hardening metals, heat can be applied where it is needed, at the temperature required, and for the required length of time. The surfaces of metal parts, such as crankshafts, rolls, pins, gears, valve stems, valve seats, and the internal surfaces of engine cylinders, can be hardened without affecting the toughness of the metal beneath the surface (see Fig. 14-4). High-frequency brazing gives a smooth, tight, and strong union of the parts. In forging, the chief advantage of high-frequency heating is in the short heating cycle.

Quenched spark-gap induction heat-treating units are made in sizes ranging from 4 to 30 kilowatts input and larger. The larger units are provided with a device that corrects the power factor to almost unity.

Courtesy Lepel Corporation.

Fig. 14-3. A 45-position rotating table for tempering hand tools.

High-frequency energy is generated in the unit and applied to the object being heat-treated through "work coils." The work coils are made of copper tubing formed into a coil having either a few turns or many turns. The work coils are connected to the outside of the unit, and they are cooled with water flowing through the copper tubing.

In the heat-treating process, the objects or machine parts that are being heat treated are placed within the turns of the work coil; thus they are exposed to the field of the high-frequency current, producing the heat energy required to heat the object or a particu-

379

Courtesy Lepel Corporation.

Fig. 14-4. A two-position station and solid-state generator for hardening
    hand tools.

lar section where heating is required. Heat energy is induced only
in the area of the object that lies within the turns of the work coil.
Since the heating action occurs with extreme rapidity (almost
instantaneously), it is relatively simple to apply the work coils to a
single section of a machine part, or simultaneously to two or more
sections of the part, and to heat those sections in such a manner that
distortions or structural changes do not occur in any other section
of the heated part.

High-frequency current concentrates the heat that it induces on
the outer surface of the machine part being heated—the higher the
frequency, the more pronounced is the "skin effect." Another
advantage that can be obtained with high-frequency current is
that the work coils are not required to be fitted closely around the
object being treated. This is an advantage because the same coil
can be used to heat several different objects of different shapes

and sizes; therefore, odd-shaped coils are not necessary for heating irregular-shaped objects.

The smaller the work, the thinner the section, or the shallower the depth to be hardened, the higher will be the frequency required. A short heating period to prevent overheating adjacent areas may require a high power concentration.

Induction heating of internal surfaces can be done by using coils shaped to match the cross section of the opening. For long holes or if the power available is not sufficient, a short coil, often only one turn, is used and either the coil or the work is moved so that the heated zone passes progressively from one end to the other of the hole. For a small hole, a hairpin-shaped coil is often used, and the work is rotated to ensure even heating.

After induction heating, the work needs to be quenched in order to be hardened. Quenching may be done by immersion, usually in oil, by a liquid spray, usually water, or by self-quenching. Self-quenching is limited to special cases where the section heated is small or thin and the mass of the rest of the piece is great enough to cool the heated section. Quenching by immersion offers the advantage of even cooling and is necessary for through-heated parts. Spray quenching is most satisfactory for surface hardening. The quenching unit is often adjacent to the coil, and the quenching cycle can follow the heating cycle without removing the work from the holding fixture. Automatic timing is often applied to both the heating and quenching cycles. This allows the exact degree of hardness to be secured.

Before operating the high-frequency unit, first determine the required temperature for the particular heat-treating operation involved and the time cycle required to produce the correct temperature. If manual control of the unit is to be used, the procedure is as follows: (1) place the metal part inside the work coil; (2) step on the floor-control pedal to switch on the current and apply heat to the metal part almost instantly; (3) maintain current for the required number of seconds; (4) switch off the current; (5) remove the part; and (6) quench the part to complete the operation. An automatic timing device can be used to take the place of manual control by the operator. Water, brine, oil, or other medium can be used as a quenching medium.

Most of the advantages of induction heating and hardening have been claimed for the induction hardening of gear teeth. Spur gears are the easiest to harden by induction heating. The gear is usually placed inside a circular coil, which is combined with a quenching ring. The process is controlled by an automatic timer. During the heating cycle the gear is usually rotated at 25 to 35 r/min to ensure uniform heating. With bevel gears, the coil must conform to the face angle of the gear. Some spiral bevel gears tend to heat one side of the tooth more than the other. Often this can be overcome by applying slightly more heat to ensure the hardening of both sides.

Where the workpiece is too long to permit heating in a fixed position, progressive heating may be used. A rod or tube of steel may be fed through a heating coil so that the heat zone travels progressively along the entire length of the workpiece. The quenching ring can be placed next to the coil so the entire operation is automatic. Gear rack teeth are often hardened in this fashion.

SUMMARY

Such operations as annealing, hardening, brazing, soldering, and melting can be performed efficiently with induction-heating equipment. Almost all objects that are electrical conductors can be heated if they are exposed to a strong electromagnetic field.

In general, metals are good conductors of electricity, and their surface resistance is of such a nature that an appreciably high current can flow at the surface of the workpiece. Low frequencies require a close coupling between the coil and the work for transferring energy. High frequencies permit a wider spacing between the coil and the load or work.

Higher frequencies are needed for shallow depth of hardening, for smaller work, and for thin sections. The heating coils can be shaped to fit the work being heated. Quenching may be by immersion, by spray, or by self-quenching. Automatic timing can be applied to both heating and quenching.

Many gears are induction hardened. Spur gears with automatic timing cycles present few problems. Progressive heating can be applied to long workpieces.

REVIEW QUESTIONS

1. What is high-frequency induction?
2. What operations can be performed by high-frequency induction?
3. What is quenched spark gap?
4. Explain why a high frequency is needed for surface hardening.
5. What is progressive heating?
6. What is a hertz?
7. What are the three types of induction-heating equipment?
8. What quenching mediums may be used with induction heating?
9. How is the quenching done?
10. Why is a spur gear usually rotated while being heated?

CHAPTER 15

Furnace Brazing

A number of well-established methods of fabricating and forming metal parts of assemblies are used in the manufacturing industry. Each method has certain manufacturing advantages that cannot be provided by other methods. The method usually employed is the one that provides the lowest overall cost for making and servicing a particular product, considering all factors.

Electric-furnace brazing has become well established among the various methods of forming parts and fabricating assemblies. This method can provide certain benefits that can be obtained by no other method. It has improved quality in most instances where the method has been adopted. The electric-furnace brazing process usually has been adopted to: (1) replace another method; (2) augment a former production method; or (3) produce a new product that has been developed which would be either difficult or impossible to produce by any method other than the furnace-brazing process.

Manufacturers of automobile and refrigerator parts make up a large portion of the users of electric-furnace brazing. Great strength in the joints is important to them, and they are also attracted by the tightness, uniformity, and excellent appearance of furnace-brazed subassemblies. Practically all automobiles and many brands of refrigerators contain several furnace-brazed parts. This is also true for many adding machines, accounting machines, cash registers, typewriters, radio receiving sets, and sewing machines.

BASIC PROCESS

Brazing is a metal joining process that uses a nonferrous filler with a melting point above 800°F (427°C) but below that of the base metals. The filler metal wets the base metal when molten and flows between the close-fitting metals because of capillary attraction.

Assemblies are put together with brazing metal, usually in the form of wire, applied near the joints that are to be brazed (Fig. 15-1). Then the assemblies are passed through an electric furnace in which a reducing atmosphere prevents the metals from oxidizing, frees the metals from any oxides that are present, and thus prepares the surfaces of the parts to be wetted by the molten brazing metal. When the brazing metal melts, it creeps on the surfaces of the parts and is drawn into the joints by capillary attraction to form alloys with the metals in the workpiece. The alloys solidify and develop great strength on transfer of the workpiece to a controlled-atmosphere cooling chamber; and the assemblies cool to a temperature at which it is safe for them to contact the outside air without danger of discoloration because of oxidation. The assemblies are delivered from the furnace with strong, tight joints and with clean, bright surfaces.

Electric-furnace brazing has been substituted for torch brazing, dip brazing, soft soldering or sweating, pinning, riveting, welding, machining from solid stock, casting, and forging. Generally, electric-furnace brazing has been employed to replace or augment one of the preceding methods because of certain objections. Some

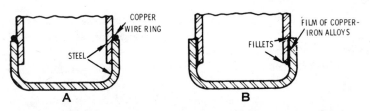

Fig. 15-1. Diagrams showing a parts assembly (A) before furnace brazing and (B) after furnace brazing.

objections in these methods that can be overcome by electric-furnace brazing are:

1. Low strength of parts; they work loose in service.
2. Nonuniform strength, which results in uncertain length of life of parts.
3. Surfaces become oxidized or covered with flux and require constant cleaning.
4. Distortions from localized heating, which require subsequent straightening or machining.
5. High cost of forming which requires machining, patterns, molds, dies, etc.
6. Low production rate which causes slow manual operations.

Another type of furnace that can be used for brazing is the salt-bath pot furnace. The heating is provided by dipping the parts into a bath of molten salt, which is heated above the melting temperature of the brazing metal.

Salt-bath brazing has four major advantages: (1) the work heats rapidly as it is submerged and in complete contact with the salt bath, (2) the salt bath protects the work from oxidation, (3) thin pieces can easily be attached to thick pieces without danger of overheating, and (4) the process can be easily adapted to a conveyor system of production work. (The reader is encouraged to review the material in Chapter 11 on pot furnaces and salt baths.)

HOLDING ASSEMBLIES TOGETHER

The relation of parts in assemblies that are to be brazed must be maintained from start to finish as the assemblies pass through the furnace. There are a number of ways of doing this, and a combination of several different methods may be necessary if the assembly is complicated. The success of a furnace-brazing job depends largely on the methods of holding the assemblies together, the design of the joints, and the means of applying the brazing metal.

The force of gravity is an important factor in furnace brazing. Assemblies of parts tend to fall apart when they are heated because the joints become loosened due to expansion. The brazing metal naturally tends to flow downward more than in other directions. Brazing metals will creep horizontally or in an upward direction on the surfaces of the metal; if applied in excess quantities, they will flow downward quite freely and collect at the low spots. In designing electric-furnace brazing assemblies, it is important to keep in mind the method of holding the assembly together inside the furnace and of setting it up inside the furnace to direct the flow of the brazing metal into the joints to give minimum distortion or movement of parts. Generally, cut-and-trial methods can be used to determine the proper procedure so that the brazing metal can be made to flow into all joints, leaving neat fillets, clean surrounding surfaces, and a job without distortion of parts.

Copper is the most common brazing metal, and steel is the most common parent metal. Other brazing metals and parent metals are also used, but the combination of copper and steel is the most common one.

LAYING AND PRESSING PARTS TOGETHER

The simplest method of joining two parts is to lay one part on top of the other with brazing metal either placed between them or wrapped around one of the members near the joint. In this method the weight of the upper member ensures good metal-to-metal contact. The chief disadvantage of this method is the lack of a means of indexing or keeping the parts from moving in relation to one another.

The most common method of assembling parts for electric-furnace brazing is to use a sleeve fit to keep them together. Regardless of the tightness of the sleeve fit, a means must be found to prevent slippage of the parts when they become heated in the furnace.

A shoulder can be machined on one member, particularly if the joint has a vertical axis, to prevent slippage and accomplish stability (Fig. 15-2). A punch can be used to turn up burrs in the shaft, as shown in an exaggerated manner in Fig. 15-3. This is called "staking," and can be used to lock two members together effectively. The staking method is used to retain the indexing of cam, lever, and gear assemblies on shafts or hubs. This can also be accomplished by tack welding (Fig. 15-4) and by pinning.

Fig. 15-2. A shoulder machined on one member is used to make a joint stable for the electric-furnace brazing process.

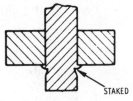

Fig. 15-3. "Staking" method used to hold parts assembly in position while it is being brazed in the electric furnace.

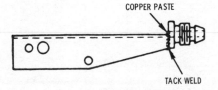

Fig. 15-4. Tack welding can be used to hold the tip to the shank of the electrode holder for brazing in the electric furnace.

Staking can be omitted where the point that is to be brazed has a horizontal axis and one part does not have a tendency to slide on the other. A snug fit is necessary. The brazing metal can be placed on the side that is more likely to draw the molten brazing metal into the joint by capillary attraction before it can flow away (Fig. 15-5). An added advantage in flow of metal can be gained by tipping the horizontal axis slightly.

When brazing steel with copper, an effort should be made to have a snug fit if possible. The usual tolerances for machined parts cannot be avoided in pressing parts together.

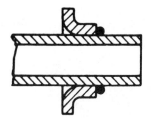

Fig. 15-5. A simple press fit can sometimes be used. The brazing metal is placed on the side that is most likely to draw the molten brazing metal into the joint by means of capillary attraction.

An inexpensive snug fit can sometimes be achieved by straight knurling the male member (Fig. 15-6). The outside diameter is considerably oversize with the hole, but wide tolerances can be used with this arrangement on both the shank and the hole, contributing toward a lower-cost job.

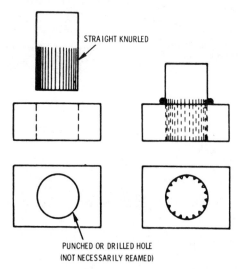

STRAIGHT KNURLED

PUNCHED OR DRILLED HOLE
(NOT NECESSARILY REAMED)

Fig. 15-6. Straight knurling on the male member can sometimes give a snug fit to hold the parts together for electric-furnace brazing.

Spot welding (Fig. 15-7) is another method used frequently to hold assembled parts for electric-furnace brazing. This is a fast and inexpensive method, and a neat job is the usual result. Spot welding prevents slippage of parts, and dimensions can be held accurately.

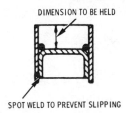

DIMENSION TO BE HELD

SPOT WELD TO PREVENT SLIPPING

Fig. 15-7. Spot welding is used frequently to hold assembled parts for electric-furnace brazing.

Parts can be "swaged" to hold them together for furnace brazing (Fig. 15-8). This is an effective and inexpensive method of assembling spuds into holes in shallow bodies. A strong, tight, and dependable bond that is leakproof and can withstand high pressures is obtained.

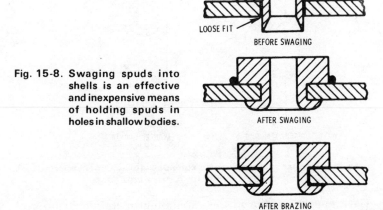

LOOSE FIT

BEFORE SWAGING

AFTER SWAGING

AFTER BRAZING

Fig. 15-8. Swaging spuds into shells is an effective and inexpensive means of holding spuds in holes in shallow bodies.

Spinning or pressing the end of a tenon into a countersunk hole can be used to lock the parts together with little or no effect on the size of the hole (Fig. 15-9). The same result can be obtained by flaring the tenon in a press. The hole is chamfered so that the end of the tenon can be spun or pressed into the chamber. This is a commonly used method for assembling parts that were formerly drilled or pinned.

PUNCHED HOLE
COUNTER SUNK

SPUN OR PRESSED

Fig. 15-9. A method of holding parts together by spinning.

When tubular members are pressed into headers, an expanded operation can be used to lock the assemblies together for furnace brazing (Fig. 15-10). Copper rings can be placed over the tubes after the expanding operation, against the outside of the tube sheet; or they can be placed over the tubes before assembly and moved against the inner surface of the tube sheet after the expanding operation, assuming that the rings are accessible.

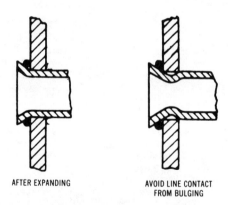

AFTER EXPANDING

AVOID LINE CONTACT
FROM BULGING

Fig. 15-10. Tubular members can be expanded into holes of headers to lock the assemblies together for furnace brazing.

Care must be taken in the expanding operation to prevent bulging of the tube to produce only a line contact (see B in Fig. 15-10). A leader on the expanding tool can be used to project into the tube to support the tube wall while the end of the tube is being flared outward.

One or more members of an assembly sometimes can be peened to hold the members together for brazing in the furnace (Fig. 15-11). Two stampings are pressed together and then peened around the edges with an air hammer. Copper in the molten state can be sprayed on the parts by means of an oxyacetylene spray gun so that it will be within the joint after assembly.

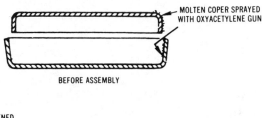

MOLTEN COPER SPRAYED WITH OXYACETYLENE GUN

BEFORE ASSEMBLY

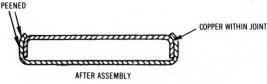

PEENED

COPPER WITHIN JOINT

AFTER ASSEMBLY

Fig. 15-11. The edges of stampings can be peened to hold snug joints as the assembly is brazed in the furnace.

A disk in the end of a shell can be held in place by crimping the ends of the shell, as illustrated in Fig. 15-12. Indentations around the shell can be used as stoppers against which the disk can be held. It is better practice to set such an assembly on end, so the brazing metal can flow downward through the joints. This is often impractical because of lengths. If the assembly must be laid on its side, an oversized ring of brazing metal can be sprung in place as near the joint as possible. A copper-powder paste can be daubed on the wire and the adjoining steel surfaces. The hardened paste tends to hold the copper wire in place so that it will not sag away from the joint at the top when it gets hot. The copper paste serves as an auxiliary supply of brazing metal and ensures the presence of brazing metal where it is needed in the joint.

Parts assemblies can also be riveted together. A fan-wheel assembly is a typical example of parts that were formerly riveted but are now riveted and then brazed in the electric furnace.

Parts can be held in their indexed positions for brazing, as shown in Fig. 15-13, by pinning. Tapered pins are driven through small levers pressed on a shaft to prevent movement during the brazing operation.

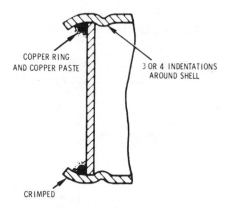

COPPER RING
AND COPPER PASTE

3 OR 4 INDENTATIONS
AROUND SHELL

CRIMPED

Fig. 15-12. Crimping the ends of a shell to hold a disk and the copper ring in position for furnace brazing.

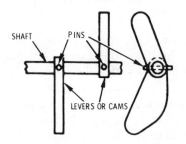

SHAFT PINS

LEVERS OR CAMS

Fig. 15-13. Tapered pins can be used to hold levers or cams in their proper indexed positions for furnace brazing.

Wedges are also used to lock parts or members into the assemblies (Fig. 15-14). Two methods of wedging are used to wedge tungsten carbide bits into the slots of milling cutters for brazing in the electric furnace. Tapered pins (A in Fig. 15-14) and wedges of mica or porcelain (b in Fig. 15-14) can be used to wedge the bits into the slots.

Screws are commonly used to hold parts together for brazing in the electric furnace (Fig. 15-15). The manner of putting the parts together is shown in the illustration.

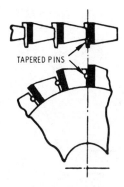

TAPERED PINS

MICA OR PORCELAIN WEDGES

Fig. 15-14. Tapered pins (left) and mica or porcelain wedges (right) can be used to wedge tungsten carbide bits into the slots of milling cutters for furnace brazing.

Fig. 15-15. Using screws to hold parts together for furnace brazing.

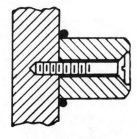

Several methods of overlapping members are effective in furnace brazing (Fig. 15-16). Copper clips can be placed at intervals over the upper edge. Thus brazing metal can be supplied to a joint that is inaccessible (*A* in Fig. 15-16).

A similar method of overlapping is sometimes used on hollow containers (*B* in Fig. 15-16). The copper can be placed inside the joint; or copper-powder paste can be applied along the outside of the seams, as indicated in the illustration.

A cross section of copper-brazed, double-walled steel tubing is made by laterally rolling a copperplated steel strip into the form shown in the illustration (*C* in Fig. 15-16). This type of tubing is used in gasoline lines of automobiles, oil lines, and hydraulic brake lines.

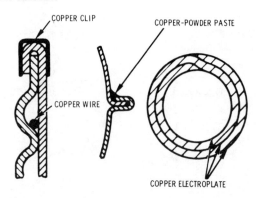

Fig. 15-16. Types of overlapping joints suitable for furnace brazing.

An undesirable type of overlapping joint that should be avoided is shown in Fig. 15-17. In long straight seams, it is almost impossible to retain surface contact within the joints because of warpage, resulting in opened or bulged joints that will leak.

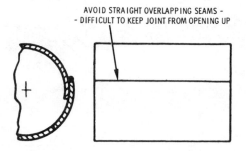

Fig. 15-17. An undesirable type of overlapping joint that should be avoided in furnace brazing because of the difficulty in keeping it tight.

Interlocked joints are used frequently in forming assemblies for furnace brazing (Fig. 15-18). The edges of steel tubing can be rolled and interlocked to prevent their opening up while passing through the furnace. A thin copperplating or wire can be used to supply brazing metal (see Fig. 15-18A). If enough time is allowed, the copper brazing metal can creep up the inner sides of the tube and braze the joint even though the seam is located at the top.

An interlocked joint that is sometimes used in making hollow tubing or hollow containers is shown in B in Fig. 15-18. This type of joint gives a secure locking effect with little possibility of its opening up when it is heated.

Some other examples of good joints for furnace brazing are shown in Fig. 15-19. Many variations of the examples given in this chapter will be found in industry, as well as other joints that require a fixture or some other method of holding the parts in place during the brazing.

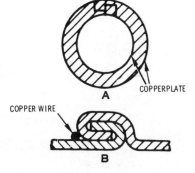

Fig. 15-18. Types of interlocking joints that are suitable for electric-furnace brazing.

SUMMARY

Electric-furnace brazing has become well established among the various methods of forming parts and fabricating assemblies. It has improved quality in most instances where the method has been adopted. Great strength in joints has been developed plus tightness, uniformity, and excellent appearance of furnace-brazing subassemblies. Practically all automobiles and many brands of refrigerators contain several furnace-brazed parts. The salt-bath pot furnace is also being used for furnace brazing.

Furnace brazing has been used as a substitute for torch brazing, dip brazing, soft soldering or sweating, pinning, riveting, welding, machining from solid stock, casting, and forging. The success of a furnace-brazing job depends largely on the method of holding the assemblies together, the design of the joints, and the means of applying the brazing metal. Copper is the most common brazing metal, and steel is the most common parent metal. Other brazing metals and parent metals are also used, but the combination of copper and steel is the most common one used.

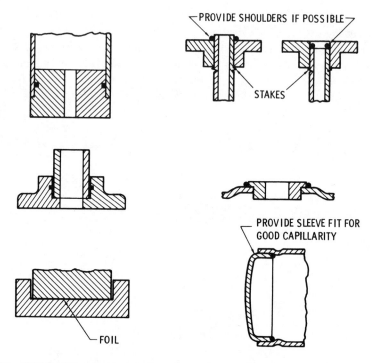

Fig. 15-19. Other examples of good joints for furnace brazing.

The simplest method of joining two parts is to lay one part on top of the other with brazing metal either placed between them or wrapped around one of the members near the joint. In this manner the weight of the upper member ensures good metal-to-metal contact. Many problems develop in the assembling of parts. One common method of assembling parts for furnace brazing is to use a sleeve fit to keep parts together. Regardless of the tightness of the sleeve fit, a means must be found to prevent slippage of parts when they become heated in the furnace.

REVIEW QUESTIONS

1. What are the advantages of furnace brazing?
2. Name the various jobs furnace brazing has eliminated?

3. What problems are encountered with assembly of various parts in furnace brazing?
4. What industries have used furnace brazing to a large degree?
5. What is the most common brazing metal? Parent metal?
6. What is brazing?
7. What two types of furnaces are used for brazing?

CHAPTER 16

Cold-Treating Process

Cold treating is a comparatively new process with many possible applications. Chilling machines are built for high-production work, and they are standard equipment in plants that use the cold-treating process.

Cold treating is used as a supplement to heat treating in the process of hardening metals. A general knowledge of both the structural changes and the essential processes of heat treatment is required to understand fully the effect of subzero temperatures on tool steels.

FUNDAMENTAL PRINCIPLE
OF COLD TREATING

In accordance with the slip-interference theory, hardness is described as a resistance to permanent deformation of the material. When either tensile or compression forces are applied to a piece of metal, the result is a slipping action along the boundaries of the grain or crystallographic planes.

The amount of slip or slide that actually occurs is dependent on the grain size, slip-interference particles, and the force brought to bear. Decreasing the size of the particles increases the number of interference particles and increases the resistance to slip, resulting in more balance.

Decalescence

As steel is heated, it passes through periods of decalescence during which structural changes occur between the base metal (iron), carbon, and the alloying elements. The result of these changes is a solid solution, which means, in this instance, that both the iron and the alloy lose their separate indentities and become a single material (*austenite*).

Rapid quenching prevents a change to the annealed condition but develops a structure called *martensite*, which has many small planes whose corners or edges project to the boundary of the grain and provide many small particles that produce maximum interference to slip, consequently developing a high degree of hardness. During this transformation stage, subzero temperatures (deep freeze) can be applied to ensure full hardness and promote uniformity of the steel.

In common heat-treating procedures, the decomposition of austenite to martensite stops when the temperature of the steel decreases to room temperature (Fig. 16-1). By cooling at temperatures of −120°F (−84°C), it is possible to complete the decomposition of the retained austenite.

Cold-Treating Temperatures

The precise subzero temperatures for cold treating depends on the metal being treated, but it has been found that metals can be cold treated between −100°F (8173°C) and −120°F (−84°C). The transformation of austenite to martensite stops at −150°F (−101°C); lower temperatures have no useful effect on high-speed steels. In addition, temperatures lower than −150°F (−101°C) can cause the metal to crack. Temperatures higher than −100°F (−73°C) are relatively ineffective in the cold-treating process.

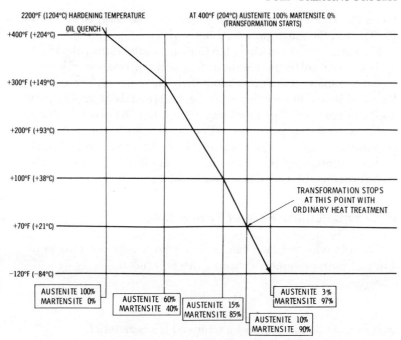

Fig. 16-1. Diagram showing the percentage of austenite transformed to martensite in high-speed steel at temperatures between 400°F (204°C) and −120°F (−84°C).

The length of time required to cold treat a metal depends on the material itself. It is necessary to chill high-speed steel thoroughly. Both ferrous and nonferrous metals, such as aluminum, cast iron, and steel, require a time cycle (alternately subjecting the parts to heat and cold, returning each time to room temperature before going to the other extreme) to stabilize the part completely. Although two to five cycles are recommended, the exact number of cycles depends on the stability required. The final cycle should be followed by temperatures of 200°F (93°C) to 300°F (149°C).

Convection Fluid

In any cold treating procedure, such as stabilizing, hardening, or shrink-fitting, heat can be dissipated more rapidly if the unit is filled to two-thirds capacity with a convection fluid, such as methylene chloride or ethyl alcohol (denatured alcohol) M.P. 179°F (82°C), and the parts immersed in the fluid.

The parts should remain in the fluid at a temperature of −120°F (−84°C) for a sufficient length of time after dissipation of heat to ensure complete transformation within the material. The precise length of time will vary; for example, 42.5 pounds of steel contain 1000 Btu per hour. The rated capacity of the Deepfreeze Model F-120 is 1000 Btu per hour. Therefore, a period of 1 hour is required to dissipate the natural heat from the 42.5 pounds of steel, and an additional 1½ to 2 hours are required to complete the transformation of austenite to martensite.

How to Calculate Rate of Production

The following formula can be used to calculate the number of Btu that must be removed from a piece of metal:

$$H = WDS$$

in which H is the heat to be removed from material;
   W is the weight of material;
   D is the degrees temperature change desired; and
   S is the specific heat of material.

Example: How many long steel bars 4¼ inches in diameter × 4 inches in length can be chilled [temperature reduced from 80°F (26.7°C) to −120°F (−84°C)] per hour?

By referring to a table of standard weights, the weight of the material can be found to be 4.02 pounds per inch, or 16.07 pounds for a 4-inch bar (4.02 × 4). The specific heat of steel, as shown in Table 16-1, is 0.118, and the desired temperature change is 200°F (93°C). Substituting in the formula ($H = WDS$):

$$H = (16.07 \times 200 \times 0.118)$$
$$= 379.252 \text{ Btu}$$

to be removed from each piece.

Since the Deepfreeze Cascade Model F-120 machine can remove 1000 Btu per hour, it is possible to handle:

$$\frac{1000}{379.252} = 2.6368 \text{ pieces per hour}$$

Table 16-1. Specific Heat of Common Materials

Aluminum	0.212	Rubber, hard	0.339
Brass	0.092	Silver	0.056
Bronze	0.104	Steel	0.118
Cast iron	0.113	Tantalum	0.033
Copper	0.092	Tin	0.054
Glass	0.180	Tungsten	0.034
Marble	0.206	Zinc	0.993
Nickel	0.109		

COLD-TREATING PROCEDURES

Several procedures can be used in cold treating the various types of steels to increase hardness by effecting a transformation of austenite to martensite. The beneficial effect of cold treating is lost unless heat treating is also properly performed in conjunction with cold treating.

High-Speed Tool Steel

Three procedures have proved satisfactory in the treatment of high-speed tool steels. It should be noted that Procedure No. 1 and Procedure No. 3 have similarities, but they differ with respect to the order in which the tempering and cold-treating operations occur. On the other hand, Procedure No 2 has two cold-treating cycles. The three procedures are as follows:

1. *Procedure No. 1*—Preheat at 1400° (760°C) to 1500°F (816°C), depending on analysis [double preheating is recommended, using 700° to 1000°F (371° to 538°C) for the first preheating].

   a. Heat to hardening temperatures, depending on analysis [Test indicates unsatisfactory results of subzero treatment on high-speed steel if the hardening temperature is near the lower part of the range. Effective temperatures appear to range from 2300° to 2350°F (1260° to 1288°C)].

405

b. Quench in oil, lead, salt, or air.

c. Remove from quenching medium at approximately 200°F (93°C) and transfer to tempering temperature.

d. Temper 2 to 4 hours to hardness specification [1000°F, (538°C) minimum].

e. Allow tool to cool to 150°F (66°C).

f. Cold treat to −120°F (−84°C) in Deepfreeze chilling machine for 3 to 6 hours, depending on cross section of material.

g. Allow tool to return to room temperature normally.

h. Repeat tempering cycle, using 25° lower temperature for 2 to 4 hours.

2. *Procedure No. 2*—Preheat as in first procedure to 1400° to 1550°F (760° to 843°C) (double preheat whenever possible).

a. Heat to hardening temperatures of 2300° to 2350°F (1260° to 1288°C), depending on analysis.

b. Quench in oil, lead, salt, or air.

c. Tool can cool in air if removed from quenching mediums at higher temperatures to 150°F (66°C).

d. Transfer to Deepfreeze unit at −120°F (−84°C) for 3 to 6 hours, depending on cross section.

e. Allow to return to room temperature.

f. Temper to specified hardness.

g. Transfer to Deepfreeze unit at −120°F(−84°C) when tool has cooled at approximately 150°F (66°C) from the tempering temperature.

h. Remove from Deepfreeze unit and allow tool to return to room temperature.

i. Retemper at 25° lower than original tempering temperature for 2 to 4 hours.

3. *Procedure No. 3*—This is the same as Procedure No. 2 except that the second subzero cooling is omitted; the cycle consists of hardening, subzero cooling, and double tempering. Note that Procedure No. 1 has only one subzero cooling, but it follows the first tempering operation.

Sometimes cutters and tools that have been heat treated in the usual manner (without subzero cooling) and finished by grinding can be improved by being subjected to the subzero treatment. Hardening and tempering tends to stabilize whatever austenite is retained; therefore, the subzero treatment of stock tools will not increase the tool durability to the same degree as when this treatment is included in the heat-treating process. Double-tempered tools show even less improvement with this after-finish subzero treatment.

If precise dimensional qualities are desired, the tool should be subjected to subzero temperature before it is finish ground. The reason for doing this is that in accomplishing the transformation of the retained austenite to martensite, a slight increase in the size of the tool results because martensite is larger in volume than austenite. Therefore, if finished stock tools are taken from the tool crib and subjected to a subzero treatment, those requiring absolute dimensional stability should be measured both before and after treatment to determine whether the original dimensions have changed.

If a slight increase in the size of the tool results from cold treating, the tool can usually be returned to correct size by grinding. Hardness of cutting tools can increase perceptibly with this treatment if the original tempering operation was incomplete. If this increase in hardness does occur, the tools can be drawn back to the usual hardness working ranges.

High-Carbon Steel

The amount of austenite retained in ordinary carbon steel is usually not large enough to be detrimental; but in some instances where the machine part is to be placed in extremely hard usage, it is necessary to complete the transformation of austenite by means of subzero temperatures. The cupping punch that is used in drawing out steel cartridges is a typical example. Formerly, only 3000 shells could be drawn out with a punch; but after cold treating, 30,000 shells could be drawn with a punch.

For another example, the tool life of step-cut reamers was increased from 100 to 400 holes by subjecting the reamers to a temperature of −120°F (−84°C) for a period of 2 to 4 hours. The flute ends also wear evenly rather than becoming tapered.

Stabilizing Dimensions

The transformation of austenite may occur naturally over a period of months or years. This transformation into martensite is accompanied by an increase in volume. Such changes may be serious in the case of precision gages, close-fitting machine parts, etc. The subzero treatment has proved effective in preventing such changes. Gage blocks, for example, are stabilized by hardening followed by repeated cycles (2 to 6 cycles are common) of chilling and tempering. This transforms a large percentage of the austenite into martensite.

Subzero treatment will always cause an increase in size. Machine parts subjected to repeated and drastic changes in temperature, such as in aircraft, may eventually cause trouble. As the austenite gradually changes to martensite, the pieces grow or warp and may cause a seizure or some other type of failure. The practical remedy is to apply the subzero treatment before the final machining operation.

SUBZERO CHILLING

Subzero chilling, another aspect of the cold-treating process, is used in industry to (1) shrink metals for fitting and assembling and (2) test aircraft instruments and materials.

One of the chief advantages of chilling parts for shrink-fit assembly is that there is less possibility of altering the characteristics of the metal by chilling than by heating. Tools or parts that have sections or portions that differ in size are less likely to be distorted when the chilling method is used. Chilled parts are also easier to handle on the assembly floor than heated parts. The hazard of oxidation is eliminated on chilled parts, and a finishing operation is not required to remove the oxidation from the surface of the metal.

Subzero chilling is used extensively to test aircraft instruments and materials. It lends itself to testing instruments and materials that normally function in the low temperatures encountered at high altitudes by aircraft.

The use of subzero temperatures in testing aircraft parts has resulted in many improved developments in metals, plastics, rubber, and other materials that have proved invaluable in modern production. It is usually comparatively simple to adapt the chilling unit for testing these materials to subzero temperatures.

SUMMARY

A cold-treating process is used as a supplement to heat-treating in hardening metals. Cold-treating machines are built for high production work and are standard equipment in plants that use the cold-treating process.

As steel is heated, it passes through periods of decalescence during which structural changes occur between the base metals. The result of these changes is a solid solution, which means, in this instance, that both the iron and the alloy lose their separate identities and become a single material called austenite. Rapid cool-off prevents a change to the annealed condition but develops a structure called martensite.

In common heat-treating procedures, the decomposition of austenite to martensite stops when the temperature of the steel decreases to room temperature. The precise subzero temperatures for cold treating depend on the metal being treated, but it has been found that metals can be cold treated between −100°F (−73°C) and −120°F (−84°C). The transformation of austenite to martensite stops at −150°F (−101°C); lower temperatures have no useful effect on high-speed steels. In addition, temperatures lower than −150°F (−101°C) can cause metal to crack.

Several procedures can be used in cold treating the various types of steels to increase hardness. The beneficial effect of cold treating is lost unless heat treating is properly performed in conjunction with cold treating.

Subzero treatment can transform a large percentage of the austenite into martensite. This, in turn, causes the tool or machine part to become stabilized in size. Thus, the treatment is used for precision gages and parts that are subjected to large temperature changes.

Subzero chilling is also used in industry to shrink metals for fitting and assembling and to test aircraft instruments and materials. Improved development in metals, rubber, plastic, and other materials has proved invaluable through the use of subzero temperature testing.

REVIEW QUESTIONS

1. What is austenite?
2. Explain martensite.
3. At what temperature will metal generally crack?
4. What are the temperatures for the cold-treating process?
5. Explain the need for subzero treatment for precision gages.
6. Why does metal increase in size when subjected to cold treating?
7. Why is subzero chilling used in industry?
8. What products have been improved through a chilling process?

Appendix

Colors and Approximate Temperatures
for Carbon Steel

Black Red	990°F	532°C
Dark Blood Red	1050	566
Dark Cherry Red	1175	635
Medium Cherry Red	1250	677
Full Cherry Red	1375	746
Light Cherry, Scaling	1550	843
Salmon, Free Scaling	1650	899
Light Salmon	1725	946
Yellow	1825	996
Light Yellow	1975	1080
White	2220	1216

Nominal Dimensions of Hex Bolts and Hex Cap Screws

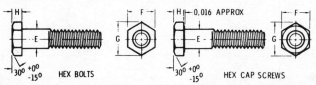

HEX BOLTS		HEX CAP SCREWS	
Nominal Size E	Width Across Flats F	Width Across Corners G	Head Height H
$^1/_4$	$^7/_{16}$	$^1/_2$	$^{11}/_{64}$
$^5/_{16}$	$^1/_2$	$^9/_{16}$	$^7/_{32}$
$^3/_8$	$^9/_{16}$	$^{21}/_{32}$	$^1/_4$
$^7/_{16}$	$^5/_8$	$^{47}/_{64}$	$^{19}/_{64}$
$^1/_2$	$^3/_4$	$^{55}/_{64}$	$^{11}/_{32}$
$^5/_8$	$^{15}/_{16}$	$1^3/_{32}$	$^{27}/_{64}$
$^3/_4$	$1^1/_8$	$1^{19}/_{64}$	$^1/_2$
$^7/_8$	$1^5/_{16}$	$1^{33}/_{64}$	$^{37}/_{64}$
1	$1^1/_2$	$1^{47}/_{64}$	$^{43}/_{64}$
$1^1/_8$	$1^{11}/_{16}$	$1^{61}/_{64}$	$^3/_4$
$1^1/_4$	$1^7/_8$	$2^{11}/_{64}$	$^{27}/_{32}$
$1^3/_8$	$2^1/_{16}$	$2^3/_8$	$^{29}/_{32}$
$1^1/_2$	$2^1/_4$	$2^{19}/_{32}$	1
$1^3/_4$	$2^5/_8$	$3^1/_{32}$	$1^5/_{32}$
2	3	$3^{15}/_{32}$	$1^{11}/_{32}$

Nominal Dimensions of Heavy Hex Bolts and Heavy Hex Cap Screws

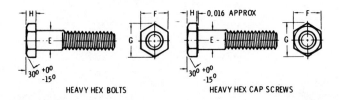

HEAVY HEX BOLTS   HEAVY HEX CAP SCREWS

HEAVY HEX BOLTS			HEAVY HEX CAP SCREWS	
	Width		**Height**	
Nominal Size	**Across Flats F**	**Across Corner G**	**Bolts H**	**Screws H**
$\frac{1}{2}$	$\frac{7}{8}$	1	$\frac{11}{32}$	$\frac{5}{16}$
$\frac{5}{8}$	$1\frac{1}{16}$	$1\frac{15}{64}$	$\frac{27}{64}$	$\frac{25}{64}$
$\frac{3}{4}$	$1\frac{1}{4}$	$1\frac{7}{16}$	$\frac{1}{2}$	$\frac{15}{32}$
$\frac{7}{8}$	$1\frac{7}{16}$	$1\frac{21}{32}$	$\frac{37}{64}$	$\frac{35}{64}$
1	$1\frac{5}{8}$	$1\frac{7}{8}$	$\frac{43}{64}$	$\frac{39}{64}$
$1\frac{1}{8}$	$1\frac{13}{16}$	$2\frac{3}{32}$	$\frac{3}{4}$	$\frac{11}{16}$
$1\frac{1}{4}$	2	$2\frac{5}{16}$	$\frac{27}{32}$	$\frac{25}{32}$
$1\frac{3}{8}$	$2\frac{3}{16}$	$2\frac{17}{32}$	$\frac{29}{32}$	$\frac{27}{32}$
$1\frac{1}{2}$	$2\frac{3}{8}$	$2\frac{3}{4}$	1	$\frac{15}{16}$
$1\frac{3}{4}$	$2\frac{3}{4}$	$3\frac{11}{64}$	$1\frac{5}{32}$	$1\frac{3}{32}$
2	$3\frac{1}{8}$	$3\frac{39}{64}$	$1\frac{11}{32}$	$1\frac{7}{32}$

Nominal Dimensions of Heavy Hex Structural Bolts

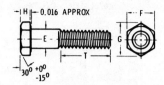

Nominal Size E	Width Across Flats F	Width Across Corners G	Head Height H	Thread Length T
$\frac{1}{2}$	$\frac{7}{8}$	1	$\frac{5}{16}$	1
$\frac{5}{8}$	$1\frac{1}{16}$	$1\frac{15}{64}$	$\frac{25}{64}$	$1\frac{1}{4}$
$\frac{3}{4}$	$1\frac{1}{4}$	$1\frac{7}{16}$	$\frac{15}{32}$	$1\frac{3}{8}$
$\frac{7}{8}$	$1\frac{7}{16}$	$1\frac{21}{32}$	$\frac{35}{64}$	$1\frac{1}{2}$
1	$1\frac{5}{8}$	$1\frac{7}{8}$	$\frac{39}{64}$	$1\frac{3}{4}$
$1\frac{1}{8}$	$1\frac{13}{16}$	$2\frac{3}{32}$	$\frac{11}{16}$	2
$1\frac{1}{4}$	2	$2\frac{5}{16}$	$\frac{25}{32}$	2
$1\frac{3}{8}$	$2\frac{3}{16}$	$2\frac{17}{32}$	$\frac{27}{32}$	$2\frac{1}{4}$
$1\frac{1}{2}$	$2\frac{3}{8}$	$2\frac{3}{4}$	$\frac{5}{16}$	$2\frac{1}{4}$

413

Nominal Dimensions of Hex Nuts, Hex Thick Nuts, and Hex Jam Nuts

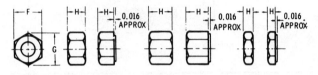

Nominal Size	Width Across Flats F	Width Across Corners G	Thickness Hex Nuts H	Thick Nuts H	Jam Nuts H
1/4	7/16	1/2	7/32	9/32	5/32
5/16	1/2	9/16	17/64	21/64	3/16
3/8	9/16	21/32	21/64	13/32	7/32
7/16	11/16	51/64	3/8	29/64	1/4
1/2	3/4	55/64	7/16	9/16	5/16
9/16	7/8	1	31/64	39/64	5/16
5/8	15/16	1 3/32	35/64	23/32	3/8
3/4	1 1/8	1 19/64	41/64	13/16	27/64
7/8	1 5/16	1 33/64	3/4	29/32	31/64
1	1 1/2	1 47/64	55/64	1	35/64
1 1/8	1 11/16	1 61/64	31/32	1 5/32	39/64
1 1/4	1 7/8	2 11/64	1 1/16	1 1/4	23/32
1 3/8	2 1/16	2 3/8	1 11/64	1 3/8	25/32
1 1/2	2 1/4	2 19/32	1 9/32	1 1/2	27/32

Nominal Dimensions of Square Head Bolts

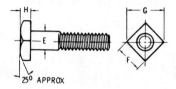

25⁰ APPROX

Nominal Size E	Width Across Flats F	Width Across Corners G	Head Height H
$1/4$	$3/8$	$17/32$	$11/64$
$5/16$	$1/2$	$45/64$	$13/64$
$3/8$	$9/16$	$51/64$	$1/4$
$7/16$	$5/8$	$57/64$	$19/64$
$1/2$	$3/4$	$1\,1/16$	$21/64$
$5/8$	$15/16$	$1\,21/64$	$27/64$
$3/4$	$1\,1/8$	$1\,19/32$	$1/2$
$7/8$	$1\,5/16$	$1\,55/64$	$19/32$
1	$1\,1/2$	$2\,1/8$	$21/32$
$1\,1/8$	$1\,11/16$	$2\,25/64$	$3/4$
$1\,1/4$	$1\,7/8$	$2\,21/32$	$27/32$
$1\,3/8$	$2\,1/16$	$2\,59/64$	$29/32$
$1\,1/2$	$2\,1/4$	$3\,3/16$	1

Nominal Dimensions of Heavy Hex Nuts
and Heavy Hex Jam Nuts

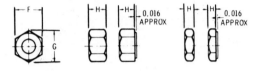

Nominal Size	Width Across Flats F	Width Across Corners G	Thickness	
			Hex Nuts H	Hex Jam Nuts H
$1/4$	$1/2$	$37/64$	$15/64$	$11/64$
$5/16$	$9/16$	$21/32$	$19/64$	$13/64$
$3/8$	$11/16$	$51/64$	$23/64$	$15/64$
$7/16$	$3/4$	$55/64$	$37/64$	$17/64$
$1/2$	$7/8$	$1\,1/64$	$31/64$	$19/64$
$9/16$	$15/16$	$1\,5/64$	$35/64$	$21/64$
$5/8$	$1\,1/16$	$1\,7/32$	$39/64$	$23/64$
$3/4$	$1\,1/4$	$1\,7/16$	$47/64$	$27/64$
$7/8$	$1\,7/16$	$1\,21/32$	$55/64$	$31/64$
1	$1\,5/8$	$1\,7/8$	$63/64$	$35/64$
$1\,1/8$	$1\,13/16$	$2\,3/32$	$1\,7/64$	$39/64$
$1\,1/4$	2	$2\,5/16$	$1\,7/32$	$23/32$
$1\,3/8$	$2\,3/16$	$2\,17/32$	$1\,11/32$	$25/32$
$1\,1/2$	$2\,3/8$	$2\,1/2$	$1\,15/32$	$27/32$
$1\,5/8$	$2\,9/16$	$2\,61/64$	$1\,19/32$	$29/32$
$1\,3/4$	$2\,3/4$	$3\,11/64$	$1\,23/32$	$31/32$
$1\,7/8$	$2\,15/16$	$3\,25/64$	$1\,27/32$	$1\,1/32$
2	$3\,1/8$	$3\,39/64$	$1\,31/32$	$1\,3/32$

Nominal Dimensions of Square Nuts
and Heavy Square Nuts

SQUARE NUTS

HEAVY SQUARE NUTS

	SQUARE NUTS		HEAVY SQUARE NUTS			
	Width Across Flats		Width Across Corners		Thickness	
Nominal Size	Regular F	Heavy F	Regular G	Heavy G	Regular H	Heavy H
$\frac{1}{4}$	$\frac{7}{16}$	$\frac{1}{2}$	$\frac{5}{8}$	$\frac{45}{64}$	$\frac{7}{32}$	$\frac{1}{4}$
$\frac{5}{16}$	$\frac{9}{16}$	$\frac{9}{16}$	$\frac{51}{64}$	$\frac{51}{64}$	$\frac{17}{64}$	$\frac{5}{16}$
$\frac{3}{8}$	$\frac{5}{8}$	$\frac{11}{16}$	$\frac{57}{64}$	$\frac{31}{32}$	$\frac{21}{64}$	$\frac{3}{8}$
$\frac{7}{16}$	$\frac{3}{4}$	$\frac{3}{4}$	$1\frac{1}{16}$	$1\frac{1}{16}$	$\frac{3}{8}$	$\frac{7}{16}$
$\frac{1}{2}$	$\frac{13}{16}$	$\frac{7}{8}$	$1\frac{5}{32}$	$1\frac{15}{64}$	$\frac{7}{16}$	$\frac{1}{2}$
$\frac{5}{8}$	1	$1\frac{1}{16}$	$1\frac{27}{64}$	$1\frac{1}{2}$	$\frac{35}{64}$	$\frac{5}{8}$
$\frac{3}{4}$	$1\frac{1}{8}$	$1\frac{1}{4}$	$1\frac{19}{32}$	$1\frac{49}{64}$	$\frac{21}{32}$	$\frac{3}{4}$
$\frac{7}{8}$	$1\frac{5}{16}$	$1\frac{7}{16}$	$1\frac{55}{64}$	$2\frac{1}{32}$	$\frac{49}{64}$	$\frac{7}{8}$
1	$1\frac{1}{2}$	$1\frac{5}{8}$	$2\frac{1}{8}$	$2\frac{19}{64}$	$\frac{7}{8}$	1
$1\frac{1}{8}$	$1\frac{11}{16}$	$1\frac{13}{16}$	$2\frac{25}{64}$	$2\frac{9}{16}$	1	$1\frac{1}{8}$
$1\frac{1}{4}$	$1\frac{7}{8}$	2	$2\frac{21}{32}$	$2\frac{53}{64}$	$1\frac{3}{32}$	$1\frac{1}{4}$
$1\frac{3}{8}$	$2\frac{1}{16}$	$2\frac{3}{16}$	$2\frac{59}{64}$	$3\frac{3}{32}$	$1\frac{13}{64}$	$1\frac{3}{8}$
$1\frac{1}{2}$	$2\frac{1}{4}$	$2\frac{3}{8}$	$3\frac{3}{16}$	$3\frac{23}{64}$	$1\frac{5}{16}$	$1\frac{1}{2}$

Nominal Dimensions of Lag Screws

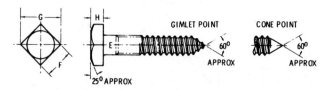

Nominal Size E	Width Across Flats F	Width Across Corners G	Head Height H
	$^9/_{32}$	$^{19}/_{64}$	$^1/_8$
$^1/_4$	$^3/_8$	$^{17}/_{32}$	$^{11}/_{64}$
$^5/_{16}$	$^1/_2$	$^{45}/_{64}$	$^{13}/_{64}$
$^3/_8$	$^9/_{16}$	$^{11}/_{16}$	$^1/_4$
$^7/_{16}$	$^5/_8$	$^{57}/_{64}$	$^{19}/_{64}$
$^1/_2$	$^3/_4$	$1^1/_{16}$	$^{21}/_{64}$
$^5/_8$	$^{15}/_{16}$	$1^{21}/_{64}$	$^{27}/_{64}$
$^3/_4$	$1^1/_8$	$1^{19}/_{32}$	$^1/_2$
$^7/_8$	$1^5/_{16}$	$1^{55}/_{64}$	$^{19}/_{32}$
1	$1^1/_2$	$2^1/_8$	$^{21}/_{32}$
$1^1/_8$	$1^{11}/_{16}$	$2^{25}/_{64}$	$^3/_4$
$1^1/_4$	$1^7/_8$	$2^{21}/_{32}$	$^{27}/_{32}$

Index

The Audel® Mail Order Bookstore

Here's an opportunity to order the valuable books you may have missed before and to build your own personal, comprehensive library of Audel books. You can choose from an extensive selection of technical guides and reference books. They will provide access to the same sources the experts use, put all the answers at your fingertips, and give you the know-how to complete even the most complicated building or repairing job, in the same professional way.

Each volume:
- **Fully illustrated**
- **Packed with up-to-date facts and figures**
- **Completely indexed for easy reference**

APPLIANCES

HOME APPLIANCE SERVICING, 3rd Edition
A practical book for electric & gas servicemen, mechanics & dealers. Covers the principles, servicing, and repairing of home appliances. 592 pages; 5½×8¼; hardbound. **Price: $12.95**

REFRIGERATION AND AIR CONDITIONING LIBRARY—2 Vols. Price: $21.95

REFRIGERATION: HOME AND COMMERCIAL
Covers the whole realm of refrigeration equipment from fractional-horsepower water coolers, through domestic refrigerators to multi-ton commercial installations. 656 pages; 5½×8¼; hardbound. **Price: $12.95**

AIR CONDITIONING: HOME AND COMMERCIAL
A concise collection of basic information, tables, and charts for those interested in understanding troubleshooting, and repairing home air conditioners and commercial installations. 464 pages; 5½×8¼; hardbound. **Price: $10.95**

OIL BURNERS, 3rd Edition
Provides complete information on all types of oil burners and associated equipment. Discusses burners—blowers—ignition transformers—electrodes—nozzles—fuel pumps—filters—controls. Installation and maintenance are stressed. 320 pages; 5½×8¼; hardbound. **Price: $9.95**

Use the order coupon on the back of this book.
All prices are subject to change without notice.

AUTOMOTIVE

AUTOMOBILE REPAIR GUIDE, 4th Edition
A practical reference for auto mechanics, servicemen, trainees, and owners. Explains theory, construction, and servicing of modern domestic motorcars. 800 pages; 5½×8¼; hardbound. **Price: $14.95**

AUTOMOTIVE AIR CONDITIONING
You can easily perform most all service procedures you've been paying for in the past. This book covers the systems built by the major manufacturers, even after-market installations. Contents: introduction—refrigerant—tools—air conditioning circuit—general service procedures—electrical systems—the cooling system—system diagnosis—electrical diagnosis—troubleshooting. 232 pages; 5½×8¼; softcover. **Price: $7.95**

DIESEL ENGINE MANUAL
A practical guide covering the theory, operation and maintenance of modern diesel engines. Explains diesel principles—valves—timing—fuel pumps—pistons and rings—cylinders—lubrication—cooling system—fuel oil and more. 480 pages; 5½×8¼; hardbound **Price: $12.95**

GAS ENGINE MANUAL, 2nd Edition
A completely practical book covering the construction, operation, and repair of all types of modern gas engines. 400 pages; 5½×8¼; hardbound. **Price: $9.95.**

BUILDING AND MAINTENANCE

ANSWERS ON BLUEPRINT READING, 3rd Edition
Covers all types of blueprint reading for mechanics and builders. This book reveals the secret language of blueprints, step by step in easy stages. 312 pages; 5½×8¼; hardbound. **Price: $9.95**

BUILDING MAINTENANCE, 2nd Edition
Covers all the practical aspects of building maintenace. Painting and decorating; plumbing and pipe fitting; carpentry; heating maintenace; custodial practices and more. (A book for building owners, managers, and maintenance personnel.) 384 pages; 5½×8¼; hardbound. **Price: $9.95**

COMPLETE BUILDING CONSTRUCTION
At last—a one volume instruction manual to show you how to construct a frame or brick building from the footings to the ridge. Build your own garage, tool shed, other outbuildings—even your own house or place of business. Building construction tells you how to lay out the building and excavation lines on the lot; how to make concrete forms and pour the footings and foundation; how to make concrete slabs, walks, and driveways; how to lay concrete block, brick and tile; how to build your own fireplace and chimney. It's one of the newest Audel books, clearly written by experts in each field and ready to help you every step of the way. 800 pages; 5½×8¼; hardbound. **Price: $19.95**

Use the order coupon on the back of this book.
All prices are subject to change without notice.

GARDENING & LANDSCAPING

A comprehensive guide for homeowners and for industrial, municipal, and estate grounds-keepers. Gives information on proper care of annual and perennial flowers; various house plants; greenhouse design and construction; insect and rodent controls; and more. 384 pages; 5½×8¼ hardbound. **Price: $9.95**

CARPENTERS & BUILDERS LIBRARY, 4th Edition (4 Vols.)

A practical, illustrated trade assistant on modern construction for carpenters, builders, and all woodworkers. Explains in practical, concise language and illustrations all the principles, advances, and shortcuts based on modern practice. How to calculate various jobs. **Price: $39.95**

> Vol. 1—Tools, steel square, saw filing, joinery cabinets. 384 pages; 5½×8¼; hardbound. **Price: $10.95**
> Vol. 2—Mathematics, plans, specifications, estimates. 304 pages; 5½×8¼; hardbound. **Price: $10.95**
> Vol. 3—House and roof framing, layout foundations. 304 pages; 5½×8¼; hardbound. **Price: $10.95**
> Vol. 4—Doors, windows, stairs, millwork, painting. 368 pages; 5½×8¼; hardbound. **Price: $10.95**

CARPENTRY AND BUILDING

Answers to the problems encountered in today's building trades. The actual questions asked of an architect by carpenters and builders are answered in this book. 448 pages; 5½×8¼; hardbound. **Price: $10.95**

WOOD STOVE HANDBOOK

The wood stove handbook shows how wood burned in a modern wood stove offers an immediate, low-cost method of full-time or part-time home heating. The book points out that wood is plentiful, low in cost (sometimes free), and nonpolluting, especially when burned in one of the newer and more efficent stoves. In this book, you will learn about the nature of heat and its control, what happens inside and outside a stove, how to have a safe and efficient chimney, and how to install a modern wood burning stove. You will learn about the differnt types of firewood and how to get it, cut it, split it and store it. 128 pages; 8½×11; softcover. **Price: $7.95**

HEATING, VENTILATING, AND AIR CONDITIONING LIBRARY (3 Vols.)

This three-volume set covers all types of furnaces, ductwork, air conditioners, heat pumps, radiant heaters, and water heaters, including swimming-pool heating systems. **Price: $38.95**

> **Volume 1**
> Partial Contents: Heating Fundamentals . . . Insulation Principles . . . Heating Fuels . . . Electric Heating System . . . Furnace Fundamentals . . . Gas-Fired Furnaces . . . Oil-Fired Furnaces . . . Coal-Fired Furnaces . . . Electric Furnances. 614 pages; 5½×8¼; hardbound. **Price: $13.95**

> **Volume 2**
> Partial Contents: Oil Burners . . . Gas Burners . . . Thermostats and Humidistats . . . Gas and Oil Controls . . . Pipes, Pipe Fitting, and Piping Details . . . Valves and Valve Installations. 560 pages; 5½×8¼; hardbound. **Price $13.95**

> **Volume 3**
> Partial Contents: Radiant Heating . . . Radiators, Convectors, and Unit Heaters . . . Stoves, Fireplaces, and Chimneys . . . Water Heaters and Other Appliances . . . Central Air Conditioning Systems . . . Humidifiers and Dehumidifiers. 544 pages; 5½×8¼; hardbound. **Price: $13.95**

HOME MAINTENANCE AND REPAIR: Walls, Ceilings, and Floors

Easy-to-follow instructions for sprucing up and repairing the walls, ceiling, and floors of your home. Covers nail pops, plaster repair, painting, paneling, ceiling and bathroom tile, and sound control. 80 pages; 8½×11; softcover. **Price: $6.95**

HOME PLUMBING HANDBOOK, 2nd Edition

A complete guide to home plumbing repair and installation. 200 pages; 8½×11; softcover. **Price: $7.95**

Use the order coupon on the back of this book.

All prices are subject to change without notice.

MASONS AND BUILDERS LIBRARY—2 Vols.

A practical, illustrated trade assistant on modern construction for bricklayers, stonemasons, cement workers, plasters, and tile setters. Explains all the principles, advances, and shortcuts based on modern practice—including how to figure and calculate various jobs. **Price $17.95**

> Vol. 1— Concrete Block, Tile, Terrazzo. 368 pages; 5½×8¼; hardbound. **Price: $9.95**
> Vol. 2—Bricklaying, Plastering, Rock Masonry, Clay Tile. 384 pages; 5½×8¼; hardbound. **Price: $9.95**

PLUMBERS AND PIPE FITTERS LIBRARY—3 Vols.

A practical, illustrated trade assistant and reference for master plumbers, journeymen and apprentice pipe fitters, gas fitters and helpers, builders, contractors, and engineers. Explains in simple language, illustrations, diagrams, charts, graphs, and pictures the principles of modern plumbing and pipe-fitting practices. **Price $29.95**

> Vol. 1—Materials, tools, roughing-in. 320 pages; 5½×8¼; hardbound. **Price: $10.95**
> Vol. 2—Welding, heating air-conditioning. 384 pages; 5½×8¼; hardbound. **Price: $10.95**
> Vol. 3—Water supply, drainage, calculations. 272 pages; 5½×8¼; hardbound. **Price: $10.95**

PLUMBERS HANDBOOK

A pocket manual providing reference material for plumbers and/or pipe fitters. General information sections contain data on cast-iron fittings, copper drainage fittings, plastic pipe, and repair of fixtures. 288 pages; 4×6 softcover. **Price: $9.95**

QUESTIONS AND ANSWERS FOR PLUMBERS EXAMINATIONS, 2nd Edition

Answers plumbers' questions about types of fixtures to use, size of pipe to install, design of systems, size and location of septic tank systems, and procedures used in installing material. 256 pages; 5½×8¼; softcover. **Price: $8.95**

TREE CARE MANUAL

The conscientious gardener's guide to healthy, beautiful trees. Covers planting, grafting, fertilizing, pruning, and spraying. Tells how to cope with insects, plant diseases, and environmental damage. 224 pages; 8½×11; softcover. **Price: $8.95**

UPHOSTERING

Upholstering is explained for the average householder and apprentice upholsterer. From repairing and regluing of the bare frame, to the final sewing or tacking, for antiques and most modern pieces, this book covers it all. 400 pages; 5½×8¼; hardbound. **Price: $12.95**

WOOD FURNITURE: Finishing, Refinishing, Repairing

Presents the fundamentals of furniture repair for both veneer and solid wood. Gives complete instructions on refinishing procedures, which includes stripping the old finish, sanding, selecting the finish and using wood fillers. 352 pages; 5½×8¼; hardbound. **Price: $9.95**

ELECTRICITY/ELECTRONICS

ELECTRICAL LIBRARY

If you are a student of electricity or a practicing electrician, here is a very important and helpful library you should consider owning. You can learn the basics of electricity, study electric motors and wiring diagrams, learn how to interpret the NEC, and prepare for the electrician's examination by using these books.

Use the order coupon on the back of this book.

All prices are subject to change without notice.

Electric Motors, 3rd Edition. 528 pages; 5½×8¼; hardbound. **Price: $12.95**

Guide to the 1981 National Electrical Code. 608 pages; 5½×8¼; hardbound. **Price: $13.95**

House Wiring, 5th Edition. 256 pages; 5½×8¼; hardbound. **Price: $9.95**

Practical Electricity, 3rd Edition. 496 pages; 5½×8¼; hardbound. **Price: $13.95**

Questions and Answers for Electricians Examinations, 7th Edition. 288 pages; 5½×8¼; hardbound. **Price: $9.95**

ELECTRICAL COURSE FOR APPRENTICES AND JOURNEYMEN
A study course for apprentice or journeymen electricians. Covers electrical theory and its applications. 448 pages; 5½×8¼; hardbound. **Price: $11.95**

RADIOMANS GUIDE, 4th Edition
Contains the latest information on radio and electronics from the basics through transistors. 480 pages; 5½×8¼; hardbound. **Price: $11.95**

TELEVISION SERVICE MANUAL, 4th Edition
Provides the practical information necessary for accurate diagnosis and repair of both black-and-white and color television receivers. 512 pages; 5½×8¼; hardbound. **Price: $11.95**

ENGINEERS/MECHANICS/ MACHINISTS

MACHINISTS LIBRARY
Covers the modern machine-shop practice. Tells how to set up and operate lathes, screw and milling machines, shapers, drill presses and all other machine tools. A complete reference library. **Price: $29.95**

Vol. 1—Basic Machine Shop. 352 pages; 5½×8¼; hardbound. **Price: $10.95**

Vol. 2—Machine Shop. 480 pages; 5½×8¼; hardbound. **Price: $10.95**

Vol. 3—Toolmakers Handy Book. 400 pages; 5½×8¼; hardbound. **Price: $10.95**

MECHANICAL TRADES POCKET MANUAL
Provides practical reference material for mechanical tradesmen. This handbook covers methods, tools equipment, procedures, and much more. 256 pages; 4×6; softcover. **Price: $8.95**

MILLWRIGHTS AND MECHANICS GUIDE
Practical information on plant installation, operation, and maintenance for millwrights, mechanics, maintenance men, erectors, riggers, foremen, inspectors, and superintendents. 960 pages; 5½×8¼; hardbound. **Price: $19.95**

POWER PLANT ENGINEERS GUIDE
The complete steam or diesel power-plant engineer's library. 816 pages; 5½×8¼; hardbound. **Price: $16.95**

QUESTIONS AND ANSWERS FOR ENGINEERS AND FIREMANS EXAMINATIONS, 3rd Edition
Presents both legitimate and "catch" questions with answers that may appear on examinations for engineers and firemans licenses for stationary, marine, and combustion engines. 496 pages; 5½×8¼; hardbound. **Price: $10.95**

WELDERS GUIDE
This new edition is a practical and concise manual on the theory, practical operation and maintenance of all welding machines. Fully covers both electric and oxy-gas welding. 928 pages; 5½×8¼; hardbound. **Price: $19.95**

Use the order coupon on the back of this book.
All prices are subject to change without notice.

WELDER/FITTERS GUIDE

Provides basic training and instruction for those wishing to become welder/fitters. Step-by-step learning sequences are presented from learning about basic tools and aids used in weldment assembly, through simple work practices, to actual fabrication of weldments. 160 pages. 8½×11; softcover. **Price: $7.95**

FLUID POWER

PNEUMATICS AND HYDRAULICS, 3rd Edition

Fully discusses installation, operation and maintenance of both HYDRAULIC AND PNEUMATIC (air) devices. 496 pages; 5½×8¼; hardbound. **Price: $10.95**

PUMPS, 3rd Edition

A detailed book on all types of pumps from the old-fashioned kitchen variety to the most modern types. Covers construction, application, installation, and troubleshooting. 480 pages; 5½×8¼; hardbound. **Price: $10.95**

HYDRAULICS FOR OFF-THE-ROAD EQUIPMENT

Everything you need to know from basic hydraulics to troubleshooting hydraulic systems on off-the-road equipment. Heavy-equipment operators, farmers. fork-lift owners and operators, mechanics—all need this practical, fully illustrated manual. 272 pages; 5½×8¼; hardbound. **Price: $8.95**

HOBBY

COMPLETE COURSE IN STAINED GLASS

Written by an outstanding artist in the field of stained glass, this book is dedicated to all who love the beauty of the art. Ten complete lessons describe the required materials, how to obtain them, and explicit directions for making several stained glass projects. 80 pages; 8½×11; softbound. **Price: $6.95**

Use the order coupon on the back of this book.
All prices are subject to change without notice.

BUILD YOUR OWN AUDEL
DO-IT-YOURSELF LIBRARY AT HOME!

Use the handy order coupon today to gain the valuable information you need in all the areas that once required a repairman. Save money and have fun while you learn to service your own air conditioner, automobile, and plumbing. Do your own professional carpentry, masonry, and wood furniture refinishing and repair. Build your own security systems. Find out how to repair your TV or Hi-Fi. Learn landscaping, upholstery, electronics and much, much more.

HERE'S HOW TO ORDER

1. Enter the correct title(s) and author(s) of the book(s) you want in the space(s) provided.

2. Print your name, address, city, state and zip code clearly.

3. Detach the order coupon below and mail today to:

Theodore Audel & Company
4300 West 62nd Street
Indianapolis, Indiana 46206
ATTENTION: ORDER DEPT.

All prices are subject to change without notice.

- -

ORDER COUPON

Please rush the following books(s).

Title _____

Author _____

Title _____

Author _____

NAME _____

ADDRESS _____

CITY _____ STATE _____ ZIP _____

☐ Payment enclosed _____
(No shipping and   Total
handling charge)

☐ Bill me (shipping and handling charge will be added)

Add local sales tax where applicable.

Litho in U.S.A.

HERE'S HOW TO ORDER

Select the Audel book(s) you want, fill in the order card below, detach and mail today. Send no money now. You'll have 15 days to examine the books in the comfort of your own home. If not completely satisfied, simply return your order and owe nothing.

If you decide to keep the books, we will bill you for the total amount, plus a small charge for shipping and handling.

1. Enter the correct title(s) and author(s) of the book(s) you want in the space(s) provided.

2. Print your name, address, city, state and zip code clearly.

3. Detach the order card below and mail today. No postage is required.

Detach postage-free order card on perforated line

FREE TRIAL ORDER CARD

☐ Please rush the following book(s) for my free trial. I understand if I'm not completely satisfied, I may return my order within 15 days and owe nothing. Otherwise, you will bill me for the total amount plus a small postage & handling charge.

Title_____

Author_____

Title_____

Author_____

NAME_____

ADDRESS_____

CITY_____ STATE_____ ZIP_____

☐ Save postage & handling costs. Full payment enclosed (plus sales tax, if any).

Cash must accompany orders under $5.00.
Money-back guarantee still applies.

DETACH POSTAGE-PAID REPLY CARD BELOW AND MAIL TODAY!

Just select your books, enter the titles and authors on the order card, fill out your name and address, and mail. There's no need to send money.

15-Day Free Trial On All Books . . .

NO POSTAGE
NECESSARY
IF MAILED
IN THE
UNITED STATES